Catalysts for the Controlled Polymerization of Conjugated Dienes

Catalysts for the Controlled Polymerization of Conjugated Dienes

Special Issue Editor

Marc Visseaux

MDPI • Basel • Beijing • Wuhan • Barcelona • Belgrade • Manchester • Tokyo • Cluj • Tianjin

Special Issue Editor
Marc Visseaux
Université des Sciences et Technologies de Lille
France

Editorial Office
MDPI
St. Alban-Anlage 66
4052 Basel, Switzerland

This is a reprint of articles from the Special Issue published online in the open access journal *Catalysts* (ISSN 2073-4344) (available at: https://www.mdpi.com/journal/catalysts/special_issues/Conjugated_Dienes).

For citation purposes, cite each article independently as indicated on the article page online and as indicated below:

LastName, A.A.; LastName, B.B.; LastName, C.C. Article Title. *Journal Name* **Year**, *Article Number*, Page Range.

ISBN 978-3-03936-190-8 (Pbk)
ISBN 978-3-03936-191-5 (PDF)

Cover image courtesy of Marc Visseaux.

Contents

About the Special Issue Editor

Marc Visseaux (Prof. Dr) (Université Lille, Pr 1C) is the team leader of MOCAH (Methodology in Organometallic Chemistry for Homogeneous Catalysis") of the Catalysis and Molecular Chemistry axis of the UCCS laboratory. After his PhD thesis in 1992 (Dijon, France), he obtained his HDR in 2000 and was appointed Full Professor in 2003, in Lille, where he joined the group of André Mortreux. MV spent a 6-month sabbatical in 2009 as Visiting Professor on the team of Professor P.L. Arnold (School of Chemistry, University of Edinburgh, UK). MV has a strong background and knowledge (97 publications, 6 patents, h index = 30, ca. 2500 citations) in polymerization reactions (with a focus on the impact of polymer chain transfer) initiated by organometallic compounds, especially with rare earths catalysts. His research focuses on chain transfer polymerization reactions (CCTP and CSP) of 1,3-dienes and olefins and he has thoroughly studied the mechanisms of ROP of lactones. He is currently working on the following: an ANRT/Cifre project obtained in 2018, an industrial collaboration contract (VYNOVA-SAV, 2019-2022), and 3 PhD theses (10 supervised theses since 2000). In addition, he is a Pedagogical Manager for large number of student courses (100 students, IUT A) in General Chemistry. Finally, he created and teaches the new courses "Homogeneous Catalysis" ENSCL, 4A, 2019; "Polymerization Catalysis", M2, 2009; and "Industrial Organic Chemistry", IUT A, 2014.

Preface to "Catalysts for the Controlled Polymerization of Conjugated Dienes"

The polymerization of conjugated dienes is a domain of interest for both academic and industrial research. In a context in which control of the process is always improving—in terms of (but not limited to) efficiency, micro-structure, etc.—and in which environmental concerns have to be taken into consideration, the development of new catalysts remains a necessary modern challenge. This includes molecular catalysts comprising less toxic metals in, for example, single component or dual catalytic combinations. The implementation of the recent concepts of the field, such as coordinative chain transfer polymerization or chain shuttling polymerization, and application to (co-)polymerization of recently introduced bio-sourced, conjugated dienes as monomers is also of interest.

The aim of this Special Issue is, thus, to cover promising recent research and novel trends in the development and application of new catalysts for conjugated diene polymerization and copolymerization. Contributions from all areas of homogeneous/supported catalysis, based on experimental results and/or mechanistic approaches, are of interest.

Marc Visseaux
Special Issue Editor

catalysts

MDPI

Editorial

Catalysts for the Controlled Polymerization of Conjugated Dienes

Marc Visseaux

UMR 8181—UCCS—Unité de Catalyse et de Chimie du Solide, ENSCL, Centrale Lille, University Artois, University Lille, CNRS, F-59000 Lille, France; marc.visseaux@ensc-lille.fr

Received: 5 October 2018; Accepted: 7 October 2018; Published: 9 October 2018

1. Background

Since its first discovery at the beginning of the 1960s [1], the coordinative polymerization of conjugated dienes has improved continuously, performer better and better. Today, chemists know how to stereospecifically polymerize conjugated dienes, whether in 1,4-*cis*, 1,4-*trans*, or 3,4(1,2) fashion. The petro-sourced (nowadays also bio-sourced for a number of them) butadiene, isoprene, and substituted conjugated diene monomers have been the subject of a very large number of studies in this context, more recently joined by natural dienes from the terpene family such as myrcene, farnesene, and ocimene. The industry has greatly helped to improve the performances of the catalytic systems (activity/productivity, selectivity, efficiency in metal catalyst), with the aim of optimizing the preparation of synthetic polymers such as 1,4-*cis* polybutadiene, which are widely applied in the tires, rubbers, and combined styrene-based resins (ABS, HIPS). Catalysts today cover a wide set of elements among which are metals from groups 4-6, 8-10 [2,3], and rare earths [4–6], while, to date, industrial concerns are mainly dominated by four metallic elements—namely neodymium, nickel, cobalt, and titanium [7]. For the synthesis of 1,4-*cis* polybutadiene, the industry catalysts are generally based on ternary systems, with a pre-catalyst associated to an activator and an aluminum chain transfer agent. The 1,4-*trans* polydienes are rather synthesized either by means of binary catalytic systems often comprising an alkylmagnesium cocatalyst, or by combination with an aluminum derivative in the case of transition metal systems. The 3,4-polydienes do not exist in the natural state, their preparation was synthetically developed later, and they have recently been shown to be potentially useful for improving tire performance, thanks to their excellent skid resistance and their low rolling resistance [8]. Nowadays, there is a better understanding of the polymerization mechanism and involves allyl-active species, thanks in particular to the support of more and more efficient calculations methods [9–11]. Since the beginning of the 2000s, there has also been a tendency for statistical copolymerization of 1,3-dienes with olefin or styrene comonomers to produce statistical, alternating, and block copolymers [12], while access to multiblock and stereoblock copolymers is currently made possible by the innovative approaches of coordinative chain transfer polymerization [13]. A last challenge is about to be solved with the preparation of stereoregular polydienes (and their copolymers) that are also end-functionalized, thanks to the living character of the polymerization. Finally, the future will probably see the development of alternative catalysts made from non-toxic and abundant metals like iron, while an even greater interest can be expected for rare earth catalysts following the discovery of new geological resources of these elements [14].

This issue brings together several important aspects of this chemistry, which remains at the forefront of both academic and industrial research interests.

2. The Present Issue

I would like to thank all of the authors and reviewers for contributing to this special issue, which together makes a nice collection of studies. I would also like to thank the editorial team and the Editor-in-Chief for their efforts in putting this issue together.

This issue comprises six research papers (five articles and one mini-review).

In the first article, Anwander and coworkers propose a global study on the use of homoleptic rare earth tetramethylaluminates for the polymerization of conjugated dienes, including a large number of rare earths elements. Activation pathways with boron derivatives and catalytic combination with $AlEt_2Cl$, for isoprene and butadiene, result in the synthesis of highly 1,4-*cis* stereoregular polymers. This study also proposes advances in reaction mechanisms, with a difference between butadiene and isoprene. The syntheses of the pre-catalysts are optimized with the X-ray structure of two of them based on gadolinium and terbium [15].

The second contribution is from the group of Ricci. Allyl cobalt phosphine complexes were prepared and characterized, they polymerized conjugated dienes (including substituted ones) via stoichiometric MAO activation. Particular behavior was observed vs. polymerization, related in particular to the internal or external substitution of the substrate diene monomer, also depending on the molecular structure (exo-exo or exo-endo orientation of the allyl group and steric bulk of additional phosphine ligands) of the pre-catalyst. Hypotheses were given to account for these observations [16].

Ricci and coworkers also contributed with a paper where a family of new trichlorobisphosphino vanadium complexes, where the phosphines were monodentate and tertiary with variations of the phosphine substituents, were used as pre-catalysts for butadiene polymerization. When such compounds were combined with MAO and TMA-free MAO, they resulted in active catalysts. The microstructure of the polybutadienes isolated was tentatively rationalized in terms of molecular structure of the pre-catalysts (better catalytic activity with lower basicity of the phosphine ligands). However, no marked differences were noted regarding the selectivity of the polymerization. This is the first example of the use of VCl_3(bisphosphine) complexes for butadiene polymerization [17].

In the fourth article, from J.-F. Carpentier and co-workers, the copolymerization of bio-renewable β-myrcene or β-farnesene with styrene was examined using an *ansa*-neodymocene catalyst, affording two series of copolymers with high styrene content and unprecedented syndioregularity of the polystyrene sequences. The incorporation of terpene in the copolymers ranged from 5.6 to 30.8 mol % (β-myrcene) and from 2.5 to 9.8 mol % (β-farnesene), respectively. NMR spectroscopy and DSC analyses suggested that the microstructure of the copolymers consists of 1,4- and 3,4-poly(terpene) units statistically distributed along the syndiotactic polystyrene chains. The thermal properties of the copolymers are strongly dependent on the terpene content, which is controlled by the initial feed. The terpolymerization of styrene with β-myrcene in the presence of ethylene was also examined [18].

The fifth article, from the group of Tanaka, was dealing with the elaboration of stereodiblock polymer of butadiene. In the first part of the study, butadiene polymerization was achieved in both a highly *cis*- or *trans*-specific manner, by using a $Cp^*Nd(BH_4)_2(THF)_2$–Bu_2Mg-d-MMAO system (d-MMAO for trialkylaluminum-depleted modified methylaluminoxane) as an initiator. This additional Al cocatalyst was added in variable amounts at the beginning of the polymerization and the *cis-/trans-* ratio could be tuned by the quantity of d-MMAO. The absence of termination or chain transfer reaction during the polymerization, deduced from the regular increase of Mn with the polymer yield, allowed further the synthesis of stereodiblock polybutadiene. This was achieved by adding dMMAO in a second time, to a polymerization mixture firstly initiated with $Cp^*Nd(BH_4)_2(THF)_2$–Bu_2Mg [19]. The stereodiblock polybutadiene thus synthesized displayed higher *cis*-regularity of the polydiene sequence, along with a broader difference of Tm and Tg temperatures, compared with the *cis/trans*- stereodiblock polyisoprene reported previously using a Nd/Mg/Al combined catalyst system based on $Nd(BH_4)_3(THF)_3$ [20].

Finally, our group contributed under the form of a mini-review, which focuses on the recent advances on the synthesis, structure, and characterization of allyl-based rare earth organometallic complexes, with emphasis on their ability to catalyze the polymerization of non-polar monomers such as conjugated dienes, styrene, and their related copolymerization [21].

References

1. Thiele, S.K.H.; Wilson, D.R. Alternate Transition Metal Complex Based Diene Polymerization. *J. Macromol. Sci. C Polym. Rev.* **2003**, *43*, 581–628. [CrossRef]
2. Ricci, G.; Sommazzi, A.; Masi, F.; Ricci, M.; Boglia, A.; Leone, G. Well-defined transition metal complexes with phosphorus and nitrogen ligands for 1,3-dienes polymerization. *Coord. Chem. Rev.* **2010**, *254*, 661–676. [CrossRef]
3. Takeuchi, D. Stereoselective Polymerization of Conjugated Dienes. In *Encyclopedia of Polymer Science and Technology*; John Wiley & Sons: Hoboken, NJ, USA, 2013.
4. Fischbach, A.; Anwander, R. Rare-Earth Metals and Aluminum Getting Close in Ziegler-Type Organometallics. *Adv. Polym. Sci.* **2006**, *204*, 155–281.
5. Friebe, L.; Nuyken, O.; Obrecht, W. Neodymium-Based Ziegler/Natta Catalysts and their Application in Diene Polymerization. *Adv. Polym. Sci.* **2006**, *204*, 1–154.
6. Zhang, Z.; Cui, D.; Wang, B.; Liu, B.; Yang, Y. Polymerization of 1,3-Conjugated Dienes with Rare-Earth Metal Precursors. *Struct. Bond.* **2010**, *137*, 49–108.
7. Srivastava, V.K.; Maiti, M.; Basak, G.C.; Jasra, R.V. Role of catalysis in sustainable production of synthetic elastomers. *J. Chem. Sci.* **2014**, *126*, 415–427. [CrossRef]
8. Yao, C.; Liu, D.; Li, P.; Wu, C.; Li, S.; Liu, B.; Cui, D. Highly 3,4-Selective Living Polymerization of Isoprene and Copolymerization with ε-Caprolactone by an Amidino *N*-Heterocyclic Carbene Ligated Lutetium Bis(alkyl) Complex. *Organometallics* **2014**, *33*, 684–691. [CrossRef]
9. Guo, H.L.; Bi, J.F.; Wu, Q.Y.; Wang, J.Y.; Shi, W.Q.; Zhang, X.Q.; Jiang, S.C.; Wu, Z.H. In situ X-ray absorption fine structure study on the polymerization of isoprene assisted by Nd-based ternary catalysts. *RSC Adv.* **2017**, *7*, 14413–14421. [CrossRef]
10. Kang, X.; Luo, Y.; Zhou, G.; Wang, X.; Yu, X.; Hou, Z.; Qu, J. Theoretical Mechanistic Studies on the *trans*-1,4-Specific Polymerization of Isoprene Catalyzed by a Cationic La–Al Binuclear Complex. *Macromolecules* **2014**, *47*, 4596–4606. [CrossRef]
11. Kefalidis, C.E.; Castro, L.; Perrin, L.; Del Rosal, I.; Maron, L. New perspectives in organolanthanide chemistry from redox to bond metathesis: Insights from theory. *Chem. Soc. Rev.* **2016**, *45*, 2516–2543. [CrossRef] [PubMed]
12. Huang, J.; Liu, Z.; Cui, D.; Liu, X. Precisely Controlled Polymerization of Styrene and Conjugated Dienes by Group 3 Single-Site Catalysts. *ChemCatChem* **2018**, *10*, 42–61. [CrossRef]
13. Valente, A.; Mortreux, A.; Visseaux, M.; Zinck, P. Coordinative Chain Transfer Polymerization. *Chem. Rev.* **2013**, *113*, 3836–3857. [CrossRef] [PubMed]
14. Takaya, Y.; Yasukawa, K.; Kawasaki, T.; Fujinaga, K.; Ohta, J.; Usui, Y.; Nakamura, K.; Kimura, J.-I.; Chang, Q.; Hamada, M.; et al. The tremendous potential of deep-sea mud as a source of rare-earth elements. *Sci. Rep.* **2018**, *8*, 5763. [CrossRef] [PubMed]
15. Hollfelder, C.O.; Jende, L.N.; Diether, D.; Zelger, T.; Stauder, R.; Maichle-Mössmer, C.; Anwander, R. 1,3-Diene Polymerization Mediated by Homoleptic Tetramethylaluminates of the Rare-Earth Metals. *Catalysts* **2018**, *8*, 61. [CrossRef]
16. Ricci, G.; Boccia, A.C.; Leone, G.; Forni, A. Novel Allyl Cobalt Phosphine Complexes: Synthesis, Characterization and Behavior in the Polymerization of Allene and 1,3-Dienes. *Catalysts* **2017**, *7*, 381. [CrossRef]
17. Leone, G.; Zanchin, G.; Pierro, I.; Sommazzi, A.; Forni, A.; Ricci, G. Synthesis, Structure and 1,3-Butadiene Polymerization Behavior of Vanadium(III) Phosphine Complexes. *Catalysts* **2017**, *7*, 369. [CrossRef]
18. Laur, E.; Welle, A.; Vantomme, A.; Brusson, J.-M.; Carpentier, J.-F.; Kirillov, E. Stereoselective Copolymerization of Styrene with Terpenes Catalyzed by an *Ansa*-Lanthanidocene Catalyst: Access to New Syndiotactic Polystyrene-Based Materials. *Catalysts* **2017**, *7*, 361. [CrossRef]
19. Tanaka, R.; Shinto, Y.; Nakayama, Y.; Shiono, T. Synthesis of Stereodiblock Polybutadiene Using $Cp^*Nd(BH_4)_2(thf)_2$ as a Catalyst. *Catalysts* **2017**, *7*, 284. [CrossRef]

20. Tanaka, R.; Yuuya, K.; Sato, H.; Eberhardt, P.; Nakayama, Y.; Shiono, T. Synthesis of stereodiblock polyisoprene consisting of cis-1,4 and trans-1,4 sequences by using a neodymium catalyst: Change of the stereospecificity triggered by an aluminum compound. *Polym. Chem.* **2016**, *7*, 1239. [CrossRef]
21. Jothieswaran, J.; Fadlallah, S.; Bonnet, F.; Visseaux, M. Recent Advances in Rare Earth Complexes Bearing Allyl Ligands and Their Reactivity towards Conjugated Dienes and Styrene Polymerization. *Catalysts* **2017**, *7*, 378. [CrossRef]

catalysts

MDPI

Article

1,3-Diene Polymerization Mediated by Homoleptic Tetramethylaluminates of the Rare-Earth Metals

Christoph O. Hollfelder, Lars N. Jende, Dominic Diether, Theresa Zelger, Rita Stauder, Căcilia Maichle-Mössmer and Reiner Anwander *

Institut für Anorganische Chemie, Eberhard-Karls-Universität Tübingen, 72076 Tübingen, Germany; christoph.hollfelder@anorg.uni-tuebingen.de (C.O.H.); lars.jende@univ-rennes1.fr (L.N.J.); dominic.diether@anorg.uni-tuebingen.de (D.D.); theresa.zelger@student.uni-tuebingen.de (T.Z.); rita.stauder@student.uni-tuebingen.de (R.S.); caecilia.maichle-moessmer@uni-tuebingen.de (C.M.-M.)

* Correspondence: reiner.anwander@uni-tuebingen.de; Tel.: +49-7071-29-72069

Received: 30 December 2017; Accepted: 23 January 2018; Published: 3 February 2018

Abstract: During the past two decades homoleptic tetramethylaluminates of the trivalent rare-earth metals, $Ln(AlMe_4)_3$, have emerged as useful components for efficient catalyst design in the field of 1,3-diene polymerization. Previous work had focused on isoprene polymerization applying $Ln(AlMe_4)_3$ precatalysts with Ln = La, Ce, Pr, Nd, Gd and Y, in the presence of Et_2AlCl as an activator. Polymerizations employing $Ln(AlMe_4)_3$ with Ln = La, Y and Nd along with borate/borane co-catalysts $[Ph_3C][B(C_6F_5)_4]$, $[PhNMe_2H][B(C_6F_5)_4]$ and $[B(C_6F_5)_3]$ were mainly investigated for reasons of comparison with ancillary ligand-supported systems (cf. half-sandwich complexes). The present study investigates into a total of eleven rare-earth elements, namely Ln = La, Ce, Pr, Nd, Gd, Tb, Dy, Ho, Y, Er and Lu. A full overview on the polymerization behavior of $Ln(AlMe_4)_3$ in the presence of perfluorinated borate/borane cocatalysts and R_2AlCl-type activators (R = Me, Et) is provided, probing the monomers isoprene and 1,3-butadiene (and preliminary ethylene). Virtually complete *cis*-1,4-selectivities are obtained for several catalyst/cocatalyst combinations (e.g., $Gd(AlMe_4)_3/Me_2AlCl$, >99.9%). Insights into the 'black box' of active species are obtained by indirect observations via screening of pre-reaction time and cocatalyst concentration. The microstructure of the polydienes is investigated by combined $^1H/^{13}C$ NMR and ATR-IR spectroscopies. Furthermore, the reaction of $[LuMe_6(Li(thf)_x)_3]$ with $AlMe_3$ has been applied as a new strategy for the efficient synthesis of $Lu(AlMe_4)_3$. The solid-state structures of $Gd(AlMe_4)_3$ and $Tb(AlMe_4)_3$ are reported.

Keywords: lanthanide; rare-earth elements; synthetic rubber; 1,3-diene polymerization; alkyl; aluminum; tetramethylaluminate

1. Introduction

Since its discovery and development in the 1950s and 1960s, Ziegler-Natta polymerization catalysis has undergone various empirical optimizations regarding the composition of the catalyst mixtures applied [1–3]. While the actual active (bimetallic) catalysts/sites have remained elusive and are subject of ongoing research, the properties of the industrially fabricated polymer products have been tailored by choice of component concentrations and additives [1,4–6].

'Ziegler Mischkatalysatoren' gain their exceptional reactivity through the cooperativity of a transition metal component and an organoaluminum(magnesium) activator [1–5]. Industrial 1,3-diene polymerization processes also take advantage of Ziegler-type catalysts and ternary mixtures like carboxylate-based $Nd(O_2CR)_3/Et_3Al_2Cl_3/iBu_2AlH$ (1:1:8) or $Nd(O_2CR)_3/Et_3Al_2Cl_3/AliBu_3$ (1:1:30) [6] proved superior to ternary 'no-less-complex' d-transition metal-based catalyst systems in terms of activity and stereospecificity issues [6–9]. On the other hand, thermally stable homoleptic tetramethylaluminates of the rare-earth metals, $Ln(AlMe_4)_3$, feature a preset heterobimetallic arrangement, per se simplifying the assessment of structure reactivity relationships (ternary versus binary catalyst system) [4,10–13]. Especially, when applying dialkylaluminum chlorides as cocatalysts/activators, complexes $Ln(AlMe_4)_3$ display one of the closest possible modelling approaches to in situ generated Ziegler-type systems [14,15]. Previous studies on the use of homoleptic $Ln(AlMe_4)_3$ as precatalysts for isoprene polymerization have been reported on several occasions, some of them even out of the main spotlight of the respective article, so that they are easily missed [11–13,16–18]. Accordingly, Et_2AlCl has been applied as an activator for $Ln(AlMe_4)_3$ (Ln = La, Ce, Pr, Nd, Gd and Y) [11–13,16,17]. Moreover, investigations using borate and borane cocatalysts $[Ph_3C][B(C_6F_5)_4]$ (**A**), $[PhNMe_2H][B(C_6F_5)_4]$ (**B**) and $[B(C_6F_5)_3]$ (**C**) have been performed for reasons of comparison [13,18]. Herein, we present a full account of the polymerization performance of complexes $Ln(AlMe_4)_3$ giving consideration to eleven different rare-earth metals (La, Ce, Pr, Nd, Gd, Tb, Dy, Ho, Y, Er and Lu), as well as borate (**A**, **B**)/borane cocatalysts (**C**) and R_2AlCl activators (R = Me (**D**), Et (**E**)). The redox-active rare-earth elements Sm, Eu and Yb, favoring the formation of divalent alkylaluminate species [19], are not included in the present study. The active species involved in 1,3-diene polymerization reactions are investigated indirectly by screening of the polymerizations at various conditions applying $Nd(AlMe_4)_3$.

Furthermore, a new protocol for efficiently synthesizing homoleptic methylaluminates of the smaller lanthanides, e.g., $Lu(AlMe_4)_3$, is introduced and active catalyst systems derived from $Nd(AlMe_4)_3$ are initially probed for the polymerization of ethylene.

2. Results and Discussion

2.1. Catalyst Systems

2.1.1. Precatalyst Synthesis and Structural Characterization

Homoleptic tris(tetramethylaluminate)s of the rare-earth elements are routinely accessible in a two-step synthesis starting from the tetrahydrofuran (thf) adducts of the commercially available chlorides. Salt metathesis with lithium dimethylamide in THF gives ate complexes $[Ln(NMe_2)_3(LiCl)_3]$ ($\mathbf{1^{Ln}}$) which are treated subsequently with an excess of $AlMe_3$ in *n*-hexane to afford $Ln(AlMe_4)_3$ ($\mathbf{2^{Ln}}$; Ln = La [11,12], Ce [12,20], Pr [12], Nd [12,21], Gd [11], Tb (this work), Dy [22], Ho [12], Y [12,21], Er [20] and Lu [12] (this work)) envisaged along with dimeric $[Me_2Al(\mu\text{-}NMe_2)]_2$ (Scheme 1, route **I**) [4,11–13,20–23].

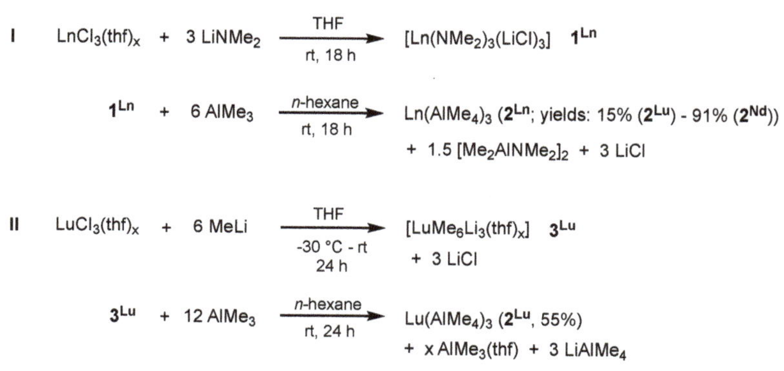

Scheme 1. Synthesis of homoleptic rare-earth metal(III) tetramethylaluminates (Ln = La ($\mathbf{2^{La}}$), Ce ($\mathbf{2^{Ce}}$), Pr ($\mathbf{2^{Pr}}$), Nd ($\mathbf{2^{Nd}}$), Gd ($\mathbf{2^{Gd}}$), Tb ($\mathbf{2^{Tb}}$), Dy ($\mathbf{2^{Dy}}$), Ho ($\mathbf{2^{Ho}}$), Y ($\mathbf{2^{Y}}$), Er ($\mathbf{2^{Er}}$) and Lu ($\mathbf{2^{Lu}}$)).

The major drawback of amide elimination protocol **I** is that the $AlMe_3$-mediated $[NMe_2] \rightarrow [AlMe_4]$ exchange in *n*-hexane provides decent yields only for the larger rare-earth metal ions [12]. In case of the smallest rare-earth metal, lutetium, purification requires subsequent sublimation allowing isolation of the desired $Lu(AlMe_4)_3$ only in ca. 15% yield [24]. Therefore, a new synthesis approach was developed, based on the trianionic hexamethylate ate complexes $[LnMe_6\{Li(Do)_x\}_3]$ (Do = tetramethylethylenediamine (tmeda), dimethoxyethane (dme), thf, diethyl ether) reported by Schumann et al. [25–27]. Treatment of $[LuMe_6\{Li(thf)_x\}_3]$ with excess of $AlMe_3$ produced the homoleptic methylaluminate complex $Lu(AlMe_4)_3$ ($\mathbf{2^{Lu}}$) in moderate crystalline yields (Scheme 1, route **II**). The side-products $LiAlMe_4$ and donor-coordinated $AlMe_3$ can easily be removed via filtration and evaporation, respectively. The absence of ate complex formation is due to the high steric saturation of the metal center by the tetramethylaluminate moieties, which show additional agostic or coordinative interactions only for the larger rare-earth metal ions [12].

The solid-state structures of complexes $\mathbf{2^{Ln}}$ employed in this study were known to all rare-earth elements except gadolinium and terbium. Putative $\mathbf{2^{Sc}}$, $\mathbf{2^{Pm}}$, and $\mathbf{2^{Eu}}$ (and hence their crystal structures) are not accessible due to reasons of stereoelectronic mismatch (Sc), radioactivity issues (Pm), and redox instability (Eu), respectively. Since complex $\mathbf{2^{Gd}}$ gave a catalyst system of exceptional performance, its crystal structure was determined (Figure 1a, Table S4.1, Supplementary Materials). To complete the series of accessible crystal structures we include also the data of $\mathbf{2^{Tb}}$ (Figure S4.1, Table S4.1, Supplementary Materials).

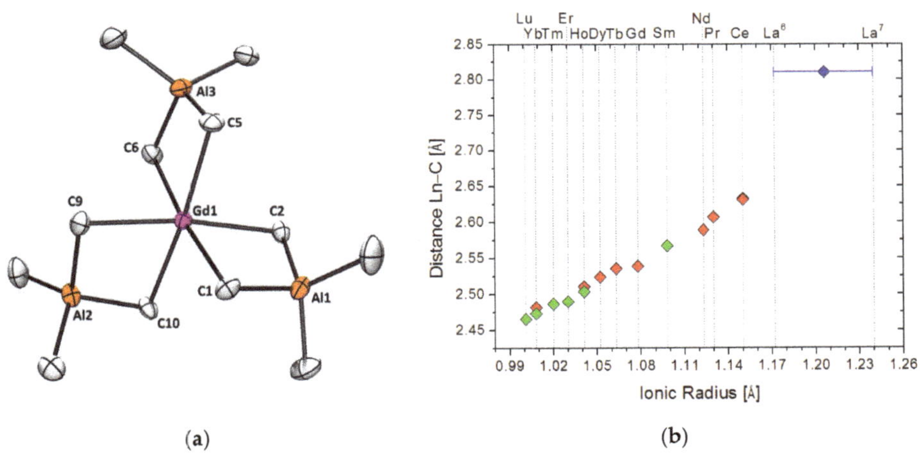

Figure 1. (**a**) ORTEP view of one of two individuals in the unit cell of $\mathbf{2^{Gd}}$. Atoms are represented by atomic displacement ellipsoids at the 50% level. Hydrogen atoms are omitted for clarity. Selected distances (Å) and angles (deg): Gd1—C1 2.543(4), Gd1—C2 2.542(4), Gd1—C5 2.550(4), Gd1—C6 2.529(4), Gd1—C9 2.529(4), Gd1—C10 2.552(4), average Gd1—CX (X = 1, 2, 5, 6, 9, 10) 2.539, Al1—C1, 2.082(4), Al1—C2, 2.086(4); C1—Gd1—C2 83.92(12), C1—Gd1—C5 92.43(13), C1—Gd1—C6 174.78(13), C1—Al1—C2 109.30(16). $\mathbf{2^{Tb}}$ crystallizes isostructurally, see Section S4, Supplementary Materials. (**b**) Overview chart on average distances of Ln—C vs. the ionic radii of the rare-earth metal trivalent cations according to Shannon and Prewitt [28,29] with a coordination number (CN) of 6. In case of $\mathbf{2^{La}}$, featuring η^3-coordination of one of the tetramethylaluminate moieties [12], the span of CN = 6 to CN = 7 is given. Crystal structures in space group $P2_1/c$ are represented by red symbols, those in $C2/c$ by green symbols and that of $\mathbf{2^{La}}$, not following the general motifs due to the coordination of one additional methyl group and crystallizing in $P2_1/n$, by a blue symbol. Data are taken from references [12,19–22] and this work.

Figure 1b gives an overview on the Ln—C distances of all known $\mathbf{2^{Ln}}$, thus displaying the first comprehensive structural data compilation on rare-earth metal alkyl complexes of the same type. While the smaller lanthanides show a tendency towards crystallization as blocks in space group $C2/c$ (green symbols) the larger representatives crystallize as needles in space group $P2_1/c$ (red symbols, two individuals per unit cell). The largest rare-earth metal center lanthanum adopts a different molecular structure (7- instead of 6-coordinate La(III) centers; space group $P2_1/n$ [12], marked in blue). The structures of $\mathbf{2^{Gd}}$ and $\mathbf{2^{Tb}}$ added to the series in this study (see Section S4, Supplementary Materials) are part of the first group and show the expected Ln—C distances to fit the linear increase with the ionic radii (Figure 1b). For Ln = Ho and Yb both crystal habits/modifications exist and were achieved by applying different crystallization parameters [19,20]. In case of Ln = Ce, two modifications in space group $P2_1/c$ are known [19,20].

2.1.2. Activation by Cationizing Cocatalysts

In order to activate the precatalysts for diene polymerization, the five most common cationizing agents were applied. Organoperfluoroborates and –borane $[Ph_3C][B(C_6F_5)_4]$ (**A**), $[PhNMe_2H][B(C_6F_5)_4]$ (**B**) and $B(C_6F_5)_3$ (**C**), respectively, cationize neutral homoleptic complexes $\mathbf{2^{Ln}}$ following ligand abstraction (**A**, **C**) and protonolysis (**B**) pathways (Scheme 2, upper part) [30–32]. Similar species were suggested to form in the silylamide-based catalyst system $Nd[N(SiMe_3)_2]_3$/**B**/Al*i*Bu_3 (1:1:10) employed for 1,3-butadiene polymerization in heptane at 70 °C (*cis*:*trans* = 86.5:11) [33].

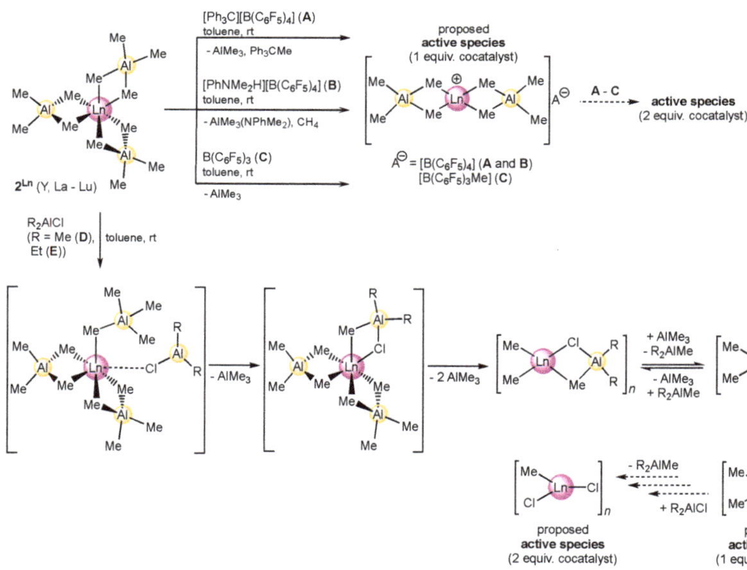

Scheme 2. Scenario of the activation of homoleptic rare-earth metal(III) tetramethylaluminates $\mathbf{2^{Ln}}$ by cocatalysts **A**–**E** (**A** = $[Ph_3C][B(C_6F_5)_4]$, **B** = $[PhNMe_2H][B(C_6F_5)_4]$, **C** = $B(C_6F_5)_3$, **D** = Me_2AlCl, **E** = Et_2AlCl), lower part adapted from ref. [10]. Activation side-product $AlMe_3(NPhMe_2)$, obtaind via activation of $\mathbf{2^{Ln}}$ with **B** is according to previous findings [31,32].

In contrast to cocatalysts **A**–**C**, R_2AlCl-based activators (**D**, R = Me; **E**, R = Et) are supposed to cationize the precatalysts by formation of large, multimetallic systems. It is presumed, that R_2AlCl replaces $AlMe_3$ in the aluminate precursors and that larger clusters form by chlorido bridging, as it has been observed for lanthanide half-sandwich complexes carrying tetramethylaluminate moieties and for lanthanidocene model systems for Ziegler-Natta catalysis [4,32,34–36], as well as for lanthanide mixed silylamide/chloride complexes [13]. With release of R_2AlMe multimetallic species are formed (Scheme 2) [4,13,35]. As the precursor carries three $[AlMe_4]^-$ moieties, the exchange of $AlMe_3$ vs. R_2AlCl can happen multiple times. Temporary re-coordination of R_2AlMe and therefore exchange of a methyl moiety in the final active species for R cannot be ruled out. This causes the presence of several distinct (cationic) clusters. As these are all assumed active species in diene polymerization with different polymerization rates, high PDI values have to be expected for the polymer products. Equilibria between clusters of different sizes could even provide enhanced complexity to these systems. It has to be mentioned, that coordination of comparably bulky substrates like monomer molecules is for all these reasons likely to have a strong impact on the structure/agglomeration of these systems. Therefore, the active initiating and propagating species might differ markedly. Interestingly, elemental analysis of the catalysts obtained from $\mathbf{2^{Nd}}$ or $\mathbf{2^{Y}}$ and **E** showed very low aluminium contents (<6%) [13], which indicates that the ratio n(Al)/n(Ln) in the active species is far smaller than 1, implying the active species being close to $[LnMe_xCl_y]_n$ (x + y = 3), shown in the lower part of Scheme 2.

2.2. Isoprene Polymerization Catalysis

2.2.1. Polymers Obtained at Standard Conditions

Previous isoprene polymerization reactions applying homoleptic complexes $\mathbf{2^{Ln}}$ were routinely run for 24 h revealing full conversion. In order to better assess the polymerization rate, in this study, a period of only 1 h was chosen as standard reaction time. Prior to monomer addition, the precatalyst and 1 or 2 equiv. of the respective cocatalyst were allowed to react for 30 min to ensure complete activation. For further details on the polymerization procedures, see Section 3.3.

Activation by a Single Equivalent of Borate/Borane. Overviewing the polymer data obtained with precatalysts $\mathbf{2^{Ln}}$ activated by borates/borane **A–C** (Figure 2a–c,f–h and Figure 3a–c; Tables S1.1.1–1.1.3, Supplementary Materials) shows high yields with both 1 and 2 equiv. of the respective cocatalyst after 1 h.

The microstructures revealed an overall increasing *cis*-content from Ce to Lu (from 44% to 77% (**A**), 44% to 78% (**B**) and, less steadily from 55% to 74% (**C**)), a decreasing *trans*-content (from 52% to 17% (**A**), 53% to 17% (**B**) and less steadily from 60% (La) to 28% (**C**)), a maximum of the 3,4-content for $\mathbf{2^{Dy}}$/**A** and $\mathbf{2^{Ho}}$/**B**, as well as a constantly levelling 3,4-content at <5% in case of cocatalyst **C** (Figure 2a–c). These findings might seem counterintuitive, as larger ions should provide more steric space for monomer coordination and the growing polymer chain and therefore favor *cis*-selectivity. As a clear identification of the active species was not successful so far, due to the paramagnetic character of most of the lanthanide ions and the low tendency of the active species toward crystallization, active species elucidation remains challenging. A reasonable interpretation of this polymerization behavior seems to be that solvent [4,5,13,37], pre-reaction side products (e.g., Ph_3CMe (**A**) and $PhNMe_2$ (**B**): Ln(III)—arene coordination) [38–40], or even anion coordination (e.g., via Ln(III)—F interactions) [41] come into play. Furthermore, dimerization might take place or a η^2-to-η^3 coordination switch of the remaining tetramethylaluminate moiety, which tendency seems more pronounced for the larger lanthanides as found for the neutral precatalysts [4,10,12]. Interestingly, the chain length/molecular weight averages of the obtained polymers were quite different throughout the series, revealing no clearly observable trend, although, a maximum in M_n seems likely for $\mathbf{2^{Dy}}$/**A** and $\mathbf{2^{Ho}}$/**B** (Figure 3a–c). Maximum PDIs were obtained for the smallest Ln(III) centers. In case of cocatalyst **C**, an increasing degree of polymerization and PDI was found with decreasing Ln ion size.

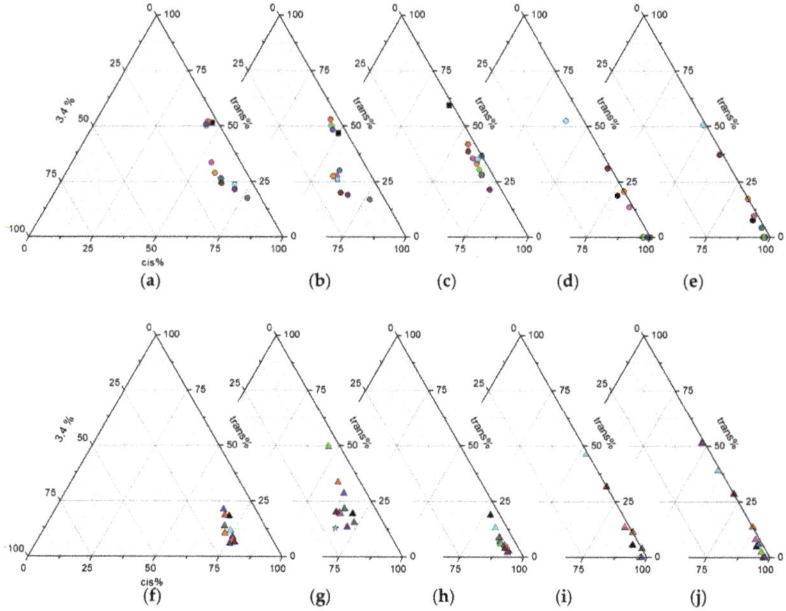

Figure 2. Ternary diagrams representing the microstructure of the polyisoprenes obtained from $Ln(AlMe_4)_3$ ($\mathbf{2}^{\mathbf{Ln}}$). Panels (**a**–**j**) give details on the selectivity obtained by the application of a certain cocatalyst (**A**–**E** (**a**–**e**) and 2 equiv. of **A**–**E** (**f**–**j**); **A** = $[Ph_3C][B(C_6F_5)_4]$, **B** = $[PhNMe_2H][B(C_6F_5)_4]$, **C** = $B(C_6F_5)_3$, **D** = Me_2AlCl, **E** = Et_2AlCl). Results for the addition of 1 equiv. of cocatalyst per $\mathbf{2}^{\mathbf{Ln}}$ after 1 h are shown by circles, 2 equiv. by triangles. If such results were not available, microstructures of reaction times of 24 h are shown marked with squares and stars, respectively. Complexes $\mathbf{2}^{\mathbf{Ln}}$ are color-coded regarding the Ln as follows: black (La), red (Ce), green (Pr), blue (Nd), teal (Gd), pink (Tb), orange (Dy), brown (Ho), cyan (Y), violet (Er) and grey (Lu). For a more traditional compilation of the polymer data in table form, sorted by precatalyst, see Tables S1.2.1—S1.2.11, Supplementary Materials.

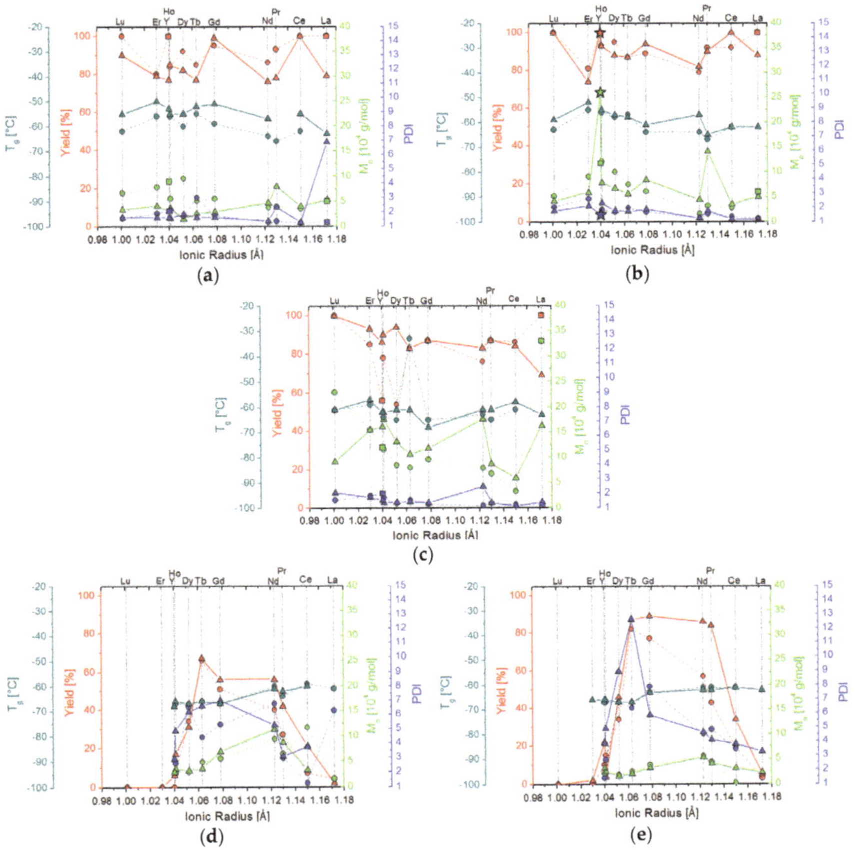

Figure 3. Overview charts representing the chain properties, yield and glass transition temperature of the polyisoprenes obtained from the homoleptic rare-earth metal tetramethylaluminates $\mathbf{2^{Ln}}$. Panels (**a**–**e**) give details on the data obtained from the application of a certain cocatalyst (**A**–**E**; **A** = $[Ph_3C][B(C_6F_5)_4]$, **B** = $[PhNMe_2H][B(C_6F_5)_4]$, **C** = $B(C_6F_5)_3$, **D** = Me_2AlCl, **E** = Et_2AlCl). Results for the addition of 1 equiv. of cocatalyst per aluminate precatalyst after 1 h are shown by circles (dashed lines), 2 equiv. by triangles (solid lines). If such results were not available, data of reaction times of 24 h are shown marked with squares and stars, respectively. The lanthanide precatalysts $Ln(AlMe_4)_3$ ($\mathbf{2^{Ln}}$) are shown in the order of increasing ionic radii (*x*-axis) [28,29]; the corresponding data are shown referring to four different y-axes: red (yield), green (M_n), blue (PDI) and teal (T_g). In all panels lines are meant to assist the reader to follow the certain curves within the plot and not to indicate linear behavior of the curves in between the points of measurement. For a more traditional compilation of the polymer data in table form, sorted by precatalyst, see Tables S1.2.1—S1.2.11, Supplementary Materials.

With respect to the uniformity of the polymerizing species (La–Lu), the ratio of the number of polymer chains and the number of lanthanide centers showed maxima with the larger Ln for **A** and **B** (e.g., 3.37 for $\mathbf{2^{Nd}}$) and minima, where M_n is at its maximum (Ln = Dy (**A**) and Ln = Ho (**B**)), while it decreased for cocatalyst **C** (with exceptions for Ln = La and Lu, Table S3.1, Supplementary Materials). Therefore, the total number of active centers was lowest for Ln = Dy (**A**) and Ln = Ho (**B**) and decreased from Ce to Er for cocatalyst **C**. This can be rationalized on the basis of coordination equilibria at least for the chain propagating species involving increasing coordination restrictions when moving from Ce to Er. For **A** and **B** then, substrate coordination capability is least pronounced, most likely caused by strong Ln^{3+}—$[B(C_6F_5)_4]^-$ interactions and competitive interactions with pre-reaction side products such as dimethylaniline. In total, these effects and peculiarities had only little influence on the glass transition temperatures, due to the vinylic content staying rather low in all cases. Besides these general trends in microstructure, several precatalyst/cocatalyst combinations stuck out. Here, especially for $\mathbf{2^{Gd}}$/**A**, a much higher *cis*-content was found, than the trends discussed (*vide infra*) would imply.

Activation by a Single Equivalent of Dialkylaluminum Chloride. Unsurprisingly, the entirely different nature of cocatalysts R_2AlCl (R = Me (**D**), Et (**E**)) and associated distinct activation pathways, led to a different set of results and interpretations (Figure 2d,e and Figure 3d,e, and Tables S1.1.4 and S1.1.5, Supplementary Materials). Most striking were the much higher *cis*-contents, achievable especially for the larger Ln ($\mathbf{2^{Nd}}$/**D**, 98.9%; $\mathbf{2^{Gd}}$/**D**, >99.9%). The *cis*-content then dropped from Ln = Ho on, giving way to an upcoming *trans*-content. As the transition state leading to *trans*-1,4-addition requires less space at the Ln(III) center than its *cis*-adding analogue, this result is in good agreement with the literature (for the evaluation of the microstructure data obtained, see Section 3.7) [42,43]. The obtained polymers showed constantly low contents of vinylic addition and therefore almost unaffected glass transition temperatures. Such binary catalyst systems $\mathbf{2^{Ln}}$/**D** and $\mathbf{2^{Ln}}$/**E** revealed a trend towards decreasing M_n for smaller Ln with (almost) no yield at all for the smallest lutetium. Interestingly, the PDI and with it the (non)uniformity of the intended multimetallic active species, revealed a maximum for Ln = Dy. These data can be interpreted by the presence of an active species, that produces 'high-*cis*'-polyisoprene for the large rare-earth metals only. Due to coordination/space restraints at smaller lanthanide ions, *cis*-selectivity decreases on the expense of increasing *trans*-contents until monomer coordination is infeasible. The high *cis*-contents reached up to regions, where the other possible microstructures could not be observed anymore at all. While in the past, this had only been found for systems applying cocatalyst **E** in *n*-hexane [11], for Ln = Gd such a high *cis*-content is also found in toluene (system $\mathbf{2^{Gd}}$/**D**, Table S1.2.5, run 7, Supplementary Materials). Since the polydiene microstructure is routinely determined by combined $^1H/^{13}C$ NMR spectroscopies and the highest *cis*-contents have been routinely observed for the rare-earth elements neodymium and gadolinium, we would like to stress a degree of uncertainty related to selectivities larger than 98%. Polymerization reactions are usually terminated by alcoholysis (=deactivation) of the catalyst system (see experimental part), resulting in inclusion of the quenching products into the polymer, which in case of, e.g., Nd(III) or Gd(III) is reflected in paramagnetic line broadening of the NMR spectra (Figure 4).

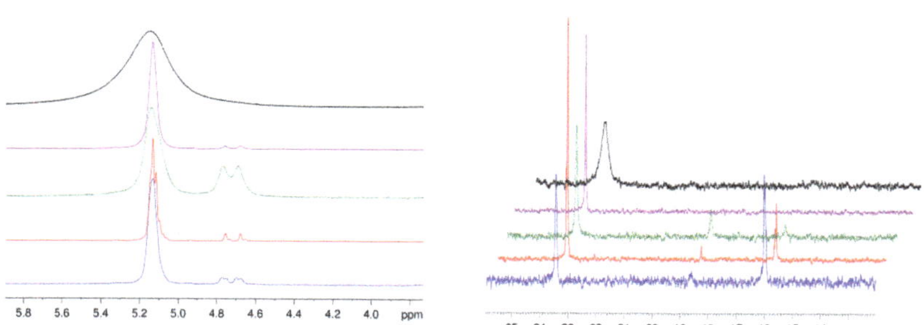

(a) (b)

Figure 4. Comparison of ^1H NMR (**a**) and ^{13}C NMR spectra (**b**) of selected polyisoprene samples (sorted by microstructure: decreasing *trans*-content bottom to top) produced by the systems $Nd(AlMe_4)_3$ ($\mathbf{2^{Nd}}$)/$[PhNMe_2H][B(C_6F_5)_4]$ (**B**) (blue; *cis*/*trans*/vinylic = 47/48/5%), $La(AlMe_4)_3$ ($\mathbf{2^{La}}$)/2 $B(C_6F_5)_3$ (**C**) (red; 78/19/4%), $Dy(AlMe_4)_3$ ($\mathbf{2^{Dy}}$)/2 $[Ph_3C][B(C_6F_5)_4]$ (**A**) (green; 72/10/18%), $Nd(AlMe_4)_3$ ($\mathbf{2^{Nd}}$)/Me_2AlCl (**D**) (violet, 99/0/1%) and $Gd(AlMe_4)_3$ ($\mathbf{2^{Gd}}$)/Me_2AlCl (**D**) (black; >99/0/0%). Inclusion of the quenching products into the polymer cannot be fully avoided. In case of polymers produced by paramagnetic catalyst systems this causes a broadening of the signals that increases the error of microstructure determination.

Moreover, the suitability of ATR-IR spectroscopy was probed as a pre-screening method for examining the microstructure of the polyisoprenes. For several polymer samples, both NMR and ATR-IR spectroscopic data have been compiled, in order to determine any significant deviation (Figure 5; Figures S2.1.1–S2.4.1 and Tables S2.1.1–S2.4.1, Supplementary Materials). Most satisfyingly, both methods are in good agreement taking into account the errors of the measurements.

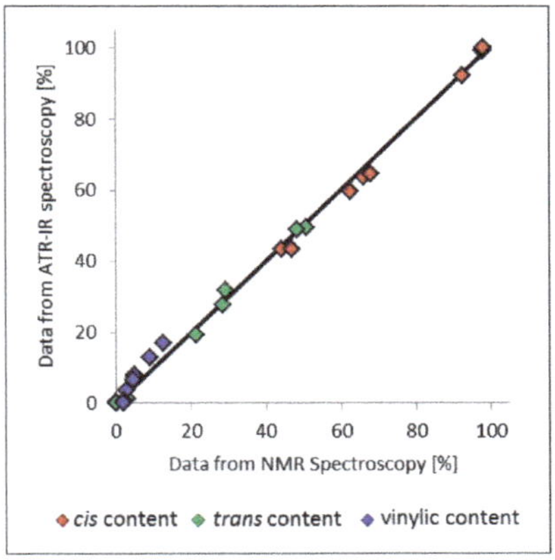

Figure 5. (See also Figure S2.2.1, Supplementary Materials). Comparison of the microstructure results obtained by NMR and ATR-IR spectroscopy, applying $Nd(AlMe_4)_3$ ($\mathbf{2^{Nd}}$) and cocatalysts **A–E** (**A** = $[Ph_3C][B(C_6F_5)_4]$, **B** = $[PhNMe_2H][B(C_6F_5)_4]$, **C** = $B(C_6F_5)_3$, **D** = Me_2AlCl, **E** = Et_2AlCl) in both 1 and 2 equiv. For detailed polymerization data see Tables S1.2.4 and S2.2.1 (Supplementary Materials).

Activation by Two Equivalents of Borate/Borane. Addition of a second equiv. of the respective cocatalyst theoretically produces two cationic charges per lanthanide metal center and is therefore assumed to change the polymerization capabilities of the active sites significantly [44,45]. Surprisingly, for cocatalysts **A** or **B** and with the exception of Ln = La, the polydispersities were found in the same range as in case of single cationization, although some polyisoprenes revealed smaller values as anticipated (Figure 3a,b, Tables S1.1.1 and S1.1.2, Supplementary Materials). Taking into account similar polymer yields, this implies that the uniformity of the active center is not much affected. In contrast, the number average molecular weights differed markedly. Although there was still no clear trend observable across the series La-Lu, polymerization initiation was visibly influenced by the second equiv. of cocatalyst as the n(chains)/n(Ln centers) ratios have changed to a large extent (Table S3.1, Supplementary Materials). Interestingly, throughout the whole Ln series, the *cis*-contents were found in the same area as observed for the catalysts derived from smaller Ln using 1 equiv. of cocatalyst. Twofold cationization is proposed to displace another tetramethylaluminate moiety, thus increasing steric unsaturation (and electron deficiency) at the rare-earth metal center. This can explain the comparatively higher *cis*-selectivities for the larger Ln, but cannot account for the concomitantly observed increased 3,4-contents (also affecting the glass transition temperatures). A preferred coordination of solvent or pre-reaction side products at such highly electron-deficient Ln(III) centers seems a plausible explanation for the latter [13,46]. When 2 equiv. of **A** were applied on the largest Ln(III), Ln = La and Pr (where Ce is an exception due to its potential to undergo redox behavior) a radical side reaction was observed that caused crosslinking of the resulting polymers and swelling on the application of solvents.

For 2 equiv. of cocatalyst **C**, even higher *cis*-contents than in case of **A** and **B** could be realized (Figure 2h, Table S1.1.3, Supplementary Materials). As with this neutral borane no pre-reaction side products are formed, competition for coordination sites with the monomer seems less pronounced, which might also explain the relatively low PDI. In contrast to the polymers formed with only 1 equiv.

of cocatalyst **C**, there is no clear trend visible for the chain length. Instead, several more or less favored combinations of ion size with its environment and monomer are suggested.

Activation by Two Equivalents of Dialkylaluminum Chloride. Addition of 2 equiv. of the cocatalysts R_2AlCl (see Section 2.2.2, for a more detailed study on the influence of the cocatalyst concentration) resulted in even more complicated systems in terms of PDI values than observed for 1:1 binary mixtures. Still, extremely high *cis*-contents and acceptable yields could be achieved for the larger Ln, down to Tb, where yields and *cis*-contents once again dropped and *trans*-1,4-addition became significant. Here, neodymium displayed the highest *cis*-contents with both cocatalysts **D** and **E** (maximum 98%). The application of R_2AlCl is limited though, as it only provides reasonable yields with the medium-sized lanthanides.

Overall Maxima. Overviewing all results (Tables S1.3.1–S1.3.7, Supplementary Materials) revealed that there are high-yielding precatalyst/cocatalyst combinations for each lanthanide and cocatalyst (cf. Figure 2). The highest *cis*-contents were found for cocatalysts R_2AlCl (**D** and **E**), especially with Ln = Ce, Pr, Nd, Gd, showing virtually complete *cis*-selectivity, while the highest induced by the borate/borane cocatalysts was found for $\mathbf{2^{Nd}}$/**2C** (92.7%). The most *trans*-selective combinations accomplished medium selectivity (<60%) for Ln = La and the borate/borane cocatalysts (**A–C**) [18]. The vinylic contents did not succeed 20% ($\mathbf{2^{Y}}$/**2B**) [16]. Interestingly, the uniformity of the active systems was highest for Ln = Ce showing even a PDI <1.2 with 2 equiv. of **B** and a surprisingly low PDI (<1.3) with 1 equiv. of **D** as well (Table S1.2.2, Supplementary Materials), in comparison with the generally high PDIs for R_2AlCl-cocatalyzed polymerizations.

2.2.2. Screening Studies

Screening of the Pre-Reaction Time. The pre-reaction time of 0.5 h has originally been chosen to ensure complete catalyst formation before monomer addition. Due to the paramagnetic character of most of the lanthanides, an exact determination of the active species by NMR spectroscopy during the process of formation remains challenging. For the few diamagnetic systems, the small total number of signals and the high mobility of the $[AlMe_4]^-$ moieties further hampered a detailed elucidation [4,10,12]. Therefore, the active species was investigated indirectly by empirically screening various pre-reaction times applying $Nd(AlMe_4)_3$ ($\mathbf{2^{Nd}}$) and cocatalyst **E**, which, in this study, served as a representative example for the R_2AlCl cocatalyst systems. It was revealed that the microstructure remained surprisingly constant (*cis*/*trans*/vinylic = 98/0/2%, Figure S1.4.1 and Table S1.4.1, Supplementary Materials). Interestingly, polymer yields were higher for shorter pre-reaction times than for the standardized 30 min, if a minimum of 10 min was allowed. At the same time, the number average molecular weight remained rather constant, which implies a higher number of initiating Nd centers under these conditions (Figure 6a). At 22 min of pre-reaction time almost every Nd center (extrapolated) produced a polymer chain (for details on the calculation of ratios n(chains)/n(Ln centers), see Section S3 in the Supplementary Materials). In addition, the ratio n(chains)/n(Ln centers) decreased below 80% at the originally applied 0.5 h. Considering the occurrence of multimetallic active species (cluster), the elucidation of the nature of such seemingly high initiation rates remain subject to future research.

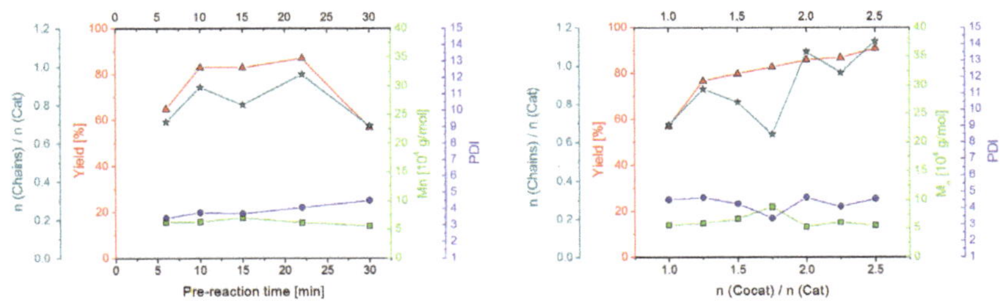

(**a**) (**b**)

Figure 6. Overview on isoprene polymerization data obtained, when investigating the influence of pre-reaction time (panel **a**) and cocatalyst concentration (**b**), applying the catalytic system $Nd(AlMe_4)_3$ ($\mathbf{2^{Nd}}$)/Et_2AlCl (**E**) with a pre-reaction time of 30 min as an example. The corresponding data are shown referring to four different *y*-axes: red (yield), green (M_n), blue (PDI) and teal (extrapolated ratio n(polymer chains)/n(Nd centers)). For details on the calculation of ratios n(chains)/n(Ln centers), see Section S3 of the Supplementary Materials. During these screenings, the microstructure remained constantly around *cis*/*trans*/3,4 = 98/0/2% (see Figure S1.4.1, Supplementary Materials). In both panels, lines are meant to assist the reader to follow the respective curves within the plot and not to indicate linear behavior of the curves in between the points of measurement.

We hypothesize that monomer coordination breaks up the multimetallic entities (compare Section 2.3.2). This theory would not interfere with the PDI values, which increase with longer pre-reaction times as a result of the formation of a variety of cationic multimetallic species. More precisely, initially formed larger units might dissociate into chain propagating species in multiple steps, releasing propagating systems stepwise and therefore subsequently. Table S3.1 (Supplementary Materials) gives an overview on the ratio n(chains)/n(Ln centers) found in this study at standard conditions (1 and 2 equiv. of cocatalyst and 0.5 h pre-reaction time, screening of precatalysts $\mathbf{2^{Ln}}$ and cocatalysts **A–E**). Apparently, values above 1, implying chain transfer at monocationic species, are more often obtained with cocatalysts **A–C**, while low values (<<1) accumulated with cocatalysts **D** and **E**. Therefore, we assume that disaggregation did not occur to full extent for all precatalysts, as the steric situations created by those vary due to the different ion sizes.

Screening of the Cocatalyst Concentration. Since multimetallic cationic species have been proposed to form upon cocatalyst addition [11], gradual variation of the cocatalyst concentration should provide further insight into the active species (Figure 6b, Table S1.4.2, Supplementary Materials). Accordingly, increasing amounts of cocatalyst resulted in an overall increase in yield and the extrapolated molar ratio of polymer chains per Nd center, while the microstructures and chain lengths remained once more rather unaffected. The ratio n(chains)/n(Ln centers) even reached beyond 1, implying minimal amounts of chain transfer. This can be explained by free $AlMe_3$, which has a low capability of serving as a chain transfer agent [47]. This would also cause the occurrence of dormant species as described by Brintzinger et al. [48]. As the capability of AlR_3 species to serve as a chain transfer agent increases with the length of R that of 'Et_2AlMe' (formed in the pre-reaction applying cocatalyst **E**) should be higher than that of $AlMe_3$ [49,50]. The slightly increasing yield with increasing cocatalyst concentration implies a slightly increasing number of catalytically active centers, in view of the almost constant molecular weight. Interpretations ascribing these findings to chain transfer behavior alone seem rather unlikely, as it would then be implausible to obtain constant number average molecular weights. Still, chain transfer plays undoubtedly a (small) role with these systems (*vide supra*). As well, it seems unlikely, that these results could be caused by a multimetallic species containing several active centers. In such a sterically crowded event, it would occur counterintuitive, that the same active species provides higher ratios n(chains)/n(Nd centers) for isoprene than for the sterically less demanding 1,3-butadiene (*vide infra*).

Isoprene Polymerization in *n*-Hexane. Polymerizations in *n*-hexane, applying $\mathbf{2^{Nd}}$/**E** (Table S1.4.3., Supplementary Materials) showed that while microstructure, molecular weight and yield are only affected slightly, the impact on the polydispersity is significant at ambient temperature. Such high PDI values can be explained by the re-dissolution of the precipitated multimetallic catalyst species upon monomer addition, which is slower in *n*-hexane than toluene [4,11,13]. The availability of the active species should be limited in a system, where the precipitate is less soluble. Increasing the solubility by application of higher temperatures (e.g., 40 °C) resulted in quantitative yields after 24 h and a lower PDI than in toluene [13]. Addition of only a small amount of toluene (Table S1.4.3, run 2) afforded the microstructure and chain length obtained in pure toluene. Interestingly, the latter system seemed to yield more lanthanide centers growing a polymer chain, which resulted in a lower PDI and higher yield.

Longer Reaction Times with Less-Active Catalysts. Precatalyst/cocatalyst combinations that have only resulted in traces of polymer after a reaction time of 1 h, were probed in 24 h terms as well (Table S1.4.4, Supplementary Materials). The binary system $\mathbf{2^{La}}$/**D** resulted in microstructure and chain properties in the anticipated range but uniquely in this study, 1,2-addition was found for one polyisoprene sample (0.8%). Furthermore, the notably high *trans*-contents observed for the smaller Ln (Er, Lu) are to mention, following the trends described in Section 2.2.1 to the maximal extent of 81% of 1,4-*trans* linkages ($\mathbf{2^{Lu}}$/2**E**).

2.3. 1,3-Butadiene Polymerization Catalysis

2.3.1. Polymers Obtained at Standard Conditions

The polymerization of 1,3-butadiene was investigated for praseodymium, neodymium and yttrium. Overall, a similar set of results was obtained as in case of isoprene polymerization although vinylic addition was even less pronounced (Figure 7a and Tables S1.6.1–S1.6.3). While the microstructure of polybutadienes obtained with $\mathbf{2^{Pr}}$ activated by borate/borane cocatalysts (**A**–**C**) showed predominant *trans*-1,4 selectivity, the high *cis*-1,4 selectivities in the presence of cocatalysts R_2AlCl (**D** and **E**) were detected for 1,3-butadiene as well. As in the isoprene case, the smaller-sized **D** produced the highest *cis*-selectivities, accomplishing 99.7% for Ln = Nd (Figure 7a, Table S1.6.2, Supplementary Materials). In all cases the yields remained rather low, probably due to shear-induced gelation of the solutions of growing polymer chains and low initiation rates. These are also the reasons for the rather high molecular weights and the variations in PDI (see Figure 7b and Tables S1.6.1–S1.6.3). For geometric reasons, 3000 equiv. of monomer per Ln center were not reached in all experiments (the actual amounts were determined by flowmeter and are given in the respective tables in Sections S1.6 and S1.7, Supplementary Materials).

2.3.2. Screening Studies

Both panels of Figure 8 show that the catalytic situation is more complicated when applying 1,3-butadiene, instead of isoprene (compare Section 2.2.2). Once again, neodymium tetramethylaluminate $\mathbf{2^{Nd}}$ was chosen as an exemplary precatalyst. The number of polymer chains produced per Nd center is significantly lower than in the isoprene case, especially at pre-reaction times longer than 15 min. This initiation deficit implies that the suggested disaggregation of the multimetallic cationic species is less pronounced (if it happens at all in the 1,3-butadiene case).

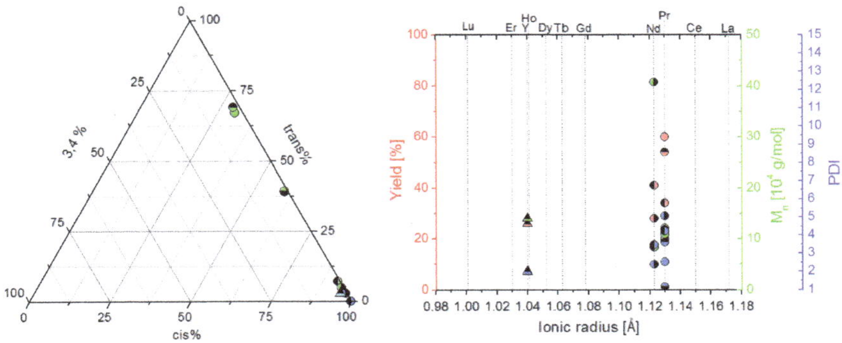

(a) (b)

Figure 7. Overview on 1,3-butadiene polymerization data. In both panels, the black parts of the marks represent the cocatalyst applied: none: **A**, top: **B**, bottom: **C**, right: **D** and left **E** (**A** = $[Ph_3C][B(C_6F_5)_4]$, **B** = $[PhNMe_2H][B(C_6F_5)_4]$, **C** = $B(C_6F_5)_3$, **D** = Me_2AlCl, **E** = Et_2AlCl). Results for the addition of 1 equiv. of cocatalyst per aluminate precatalyst are shown by circles, 2 equiv. by triangles. (**a**) ternary diagram representing the microstructure of the polybutadienes obtained from homoleptic lanthanide tetramethylaluminates. Precatalysts $\mathbf{2^{Ln}}$ (Ln = Pr, Nd, Y) are color-coded as follows: green (Pr), blue (Nd), cyan (Y). (**b**) overview chart representing the chain properties and yield of the polybutadienes. Rare-earth metal precatalysts $\mathbf{2^{Ln}}$ are shown in the order of increasing ionic radius (*x*-axis); the corresponding data are shown referring to three different *y*-axes: red (yield), green (M_n), blue (PDI).

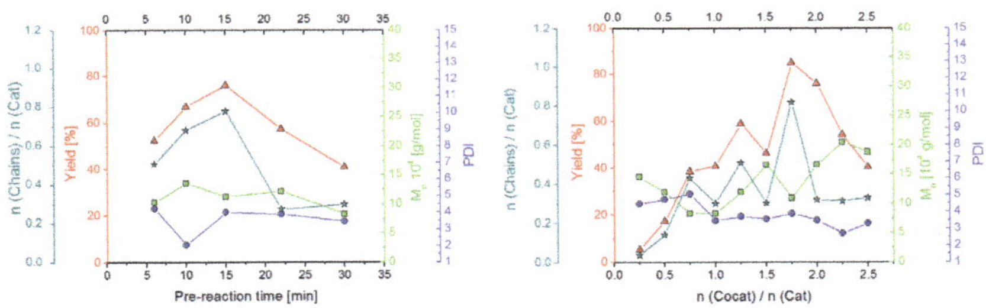

(a) (b)

Figure 8. Overview on 1,3-butadiene polymerization data obtained, when investigating the influence of pre-reaction time panel (**a**) and cocatalyst concentration (**b**), applying the catalytic system $Nd(AlMe_4)_3$ ($\mathbf{2^{Nd}}$)/Et_2AlCl (**E**) with a pre-reaction time of 30 min as an example. The corresponding data are shown referring to four different *y*-axes: red (yield), green (M_n), blue (PDI) and teal (extrapolated ratio n(polymer chains)/n(Nd centers)). For details on the calculation of ratios n(chains)/n(Ln centers), see Section S3 of the Supplementary Materials. During these screenings, the polybutadiene microstructure showed a small increase 96.7–98.8% with shorter pre-reaction times and remained constant with higher cocatalyst concentrations until 2.5 equiv. were reached, where it increased to *cis*/*trans*/vinylic = 99.5/0.4/0.1% (Figure S1.7.1 and Tables S1.7.1 and S1.7.2, Supplementary Materials). In both panels lines are meant to assist the reader to follow the certain curves within the plot and not to indicate linear behavior of the curves in between the points of measurement.

Screening of the Pre-Reaction Time. Variation of the pre-reaction time resulted in maximum monomer conversion after 15 min, while M_n remained rather constant. The latter can be explained by an increased number of active centers (or enhanced chain transfer, which once more seems unlikely, see Section 2.2.2). With an enduring activation reaction, such an increase would not be too surprising, but interestingly, after 15 min, the number of active centers seem to decrease as indicated by a significant drop of the ratio of chains per Ln center. Assuming the growth of one polymer chain per Ln(III) center, about two of three Ln centers seem not to grow a polymer chain. The following two scenarios seem likely: either it is indeed higher aggregates that propagate polymer chains or a disaggregation/re-aggregation occurs, which causes chain termination at a very early state and terminates the catalytic capabilities of certain Ln centers. While the approximate 1:2 molar ratio of polymer chains per Ln center favors the first option, the fact that PDI, degree of polymerization and even the microstructure was not affected might speak in favor of the second scenario. The highest *cis*-contents were observed at the shortest pre-reaction time (98.8% after 6 min), staying at a very high level of 96.7% even after a pre-reaction time of 30 min.

Screening of the Cocatalyst Concentration. As 30 min was the originally chosen pre-reaction time, it was also applied in the cocatalyst concentration screening (Figure 8b, Table S1.7.2, see Supplementary Materials). Not unexpectedly, using 0.25 equiv. of cocatalyst per Nd center gave low polymer yield, featuring long polymer chains as only a small number of active Nd centers propagated chain growth. Up to 0.75 equiv. of cocatalyst, the curves followed the expected paths, revealing increasing yield, decreasing molecular weight, since involving a higher number of active Nd centers. Noteworthy, a comparatively high PDI was observed for a ratio $\mathbf{E}/\mathbf{2}^{\mathbf{Nd}} < 1$. The polymerization performance appeared completely altered at higher cocatalyst concentrations. The number of active centers as well as the PDI dropped at 1 equiv. of cocatalyst. Hence, less but more uniform active centers were produced. Furthermore, the new species seemed less active as monomer uptake, that had remained constant so far, dropped as well (34.1 $\rightarrow$ 22.4 mmol), at rather constant yield (Table S1.7.2, Supplementary Materials). Despite of two exceptional catalyst/cocatalyst ratios (1.25 and 1.75), the active species present with 1 equiv. of cocatalyst seemed to stay intact up to the highest tested ratio of 2.5, as PDI and ratio n(chains)/n(Ln center) remained rather constant. As expected, cocatalyst concentrations larger than 2 equiv. resulted in lower yields due to loss of active Nd–alkyl moieties (=catalyst poisoning) [11]. Interestingly, the molar ratio of chains per Nd center seemed not affected. It is not yet clear why the catalyst species generated by 1.25 and 1.75 equiv. of cocatalyst **E** afforded a seemingly increased number of active centers and higher yield at rather unaffected polymer chain properties (M_n, PDI).

Further Studies on 1,3-Butadiene Polymerization. As a prove of principle, a 1,3-butadiene polymerization with $\mathbf{2}^{\mathbf{Nd}}/\mathbf{E}$ was run for 2 h (Table S1.7.3, Supplementary Materials). Increased yield, M_n as well as monomer uptake suggested that the reaction continues with a slight loss of polymerization rate, due to enhanced viscosity.

1,3-Butadiene polymerization with $\mathbf{2}^{\mathbf{Nd}}/\mathbf{E}$ in *n*-hexane (Table S1.7.4, Supplementary Materials) differed markedly from the same experiment performed with isoprene (Section 2.2.2). The polydispersity of the polybutadienes seemed unaffected by the solvent change. However, the number average molecular weights were much higher in *n*-hexane than in toluene at rather constant yield, suggesting that less Nd centers grow a polymer chain. The increased monomer uptake of the solution is ascribed to the different solubility of 1,3-butadiene in the applied solvents. Still, the availability of the monomer is therefore high, which causes the high M_n produced by the fewer active sites.

2.4. Ethylene Polymerization

Preliminary ethylene polymerizations have been performed with $\mathbf{2}^{\mathbf{Nd}}/\mathbf{E}$ and $\mathbf{2}^{\mathbf{Nd}}/\mathbf{2E}$ (Table S1.8.1, Supplementary Materials), affording 12% and 87% yield of polyethylene, respectively, with similar average molecular weights. The higher PDI of 4.0 in case of 2 equiv. of cocatalyst (compared to 3.4 with 1 equiv.) is reflected in the melting points of the two samples which differ by 8 K (**E**: 138 °C, **2E**: 130 °C). An extension of this study is subject to further research.

3. Materials and Methods

3.1. General Remarks

All operations were performed with rigorous exclusion of air and water, using standard Schlenk, high-vacuum and argon glovebox techniques (MBraun MB 200B; <1 ppm O_2, <1 ppm H_2O). When autoclaves did not fit into the antechambers of the applied gloveboxes (Parr type, used for ethylene polymerization), air and moisture were excluded from autoclaves as well by evacuation at reaction temperature for 24 h prior to the polymerizations and insertion of educts applying shuttles filled in a glovebox. Otherwise (Büchi type, used for 1,3-butadiene polymerization) the applied miniclaves were filled and sealed inside a glovebox. *n*-Hexane, toluene and THF (Sigma-Aldrich, St. Louis, MO, USA) were purified by using Grubbs columns (MBraun SPS-800, solvent purification system, Garching, Germany) and stored in a glovebox. C_6D_6 and toluene-d_8 were obtained from Sigma-Aldrich, dried over NaK alloy for 24 h and filtered. Anhydrous $LnCl_3$ were purchased from ABRC chemicals (Karlsruhe, Germany) and were activated by Soxhlet extraction with THF. $CDCl_3$ and $AlMe_3$ were purchased from Sigma-Aldrich and used as received. Trioctylaluminum and methyllithium were obtained from Sigma-Aldrich as solutions in hexanes and were used after solvent evaporation. $LiNMe_2$ was synthesized from $HNMe_2$ and *n*-BuLi (both Sigma-Aldrich) and employed as white powder. Isoprene was obtained from Sigma-Aldrich, dried over trioctylaluminum and vacuum transferred prior to use. $[Ph_3C][B(C_6F_5)_4]$ (**A**), $[PhNMe_2H][B(C_6F_5)_4]$ (**B**) and $B(C_6F_5)_3$ (**C**) were purchased from Boulder Scientific Company (Longmont, CO, USA) and used without any further purification. 1,3-Butadiene and ethylene, as well as nitrogen and helium (atmosphere gasses for DSC), were purchased from Westphalen Gas (Münster, Germany) and purified and dried over Grubbs columns prior to use. Their amounts were determined using a Bronkhorst EL Flow Select Flow Controller (Ruurlo, The Netherlands). Homoleptic methylaluminate complexes $\mathbf{2^{Ln}}$ were prepared according to literature methods [4,11–13,20–23]. The NMR spectra of air- and moisture-sensitive compounds were recorded at 25 °C on a Bruker DMX-400 Avance (^{1}H: 400.13 MHz; ^{13}C: 100.61 MHz; ^{27}Al: 130.33 MHz) (Billerica, MA, USA). ^{1}H, ^{13}C and ^{27}Al shifts are referenced to internal solvent resonances and reported in parts per million (ppm) relative to TMS. Elemental analyses were performed on an Elementar Vario MICRO cube (Hanau, Germany). The molar weights (M_W and M_n) of the polymers were determined by size-exclusion chromatography (SEC). Sample solutions (1.0 mg polymer per mL THF) were filtered through a 0.45 µm syringe filter (Macherey-Nagel, Düren, Germany) prior to injection. SEC was performed with a pump supplied by Viscotek (GPCmax VE 2001, Malvern, UK), employing ViscoGEL columns. Signals were detected by means of a triple detection array (TDA 305) and calibrated against polystyrene standards (M_W/M_n < 1.15; Malvern, UK, and PSS Polymer Standards Service GmbH, Mainz, Germany). The flowrate was set to 1.0 mL min^{-1}. High temperature GPC was performed by Malvern Instruments Ltd. The microstructures of the polydienes were examined by means of ^{1}H and ^{13}C NMR spectroscopic experiments on the AV400 spectrometer at ambient temperature, using $CDCl_3$ as a solvent and by ATR-IR spectroscopy on a Thermo Fisher Scientific NICOLET 6700 FTIR spectrometer using a diamond ATR setup (Waltham, MA, USA). Glass transition temperatures of the polymers (T_g) were obtained applying a Perkin-Elmer DSC 8000 calibrated with indium and cyclohexane standards, scanning from −100 °C up to +100 °C with heating rates of 20 K/min and cooling rates of 60 K/min in N_2 atmosphere (polyisoprene) and He atmosphere (polybutadiene). Crystals of $\mathbf{2^{Gd}}$ and $\mathbf{2^{Tb}}$ suitable for X-ray crystallography were grown by standard techniques from solutions using *n*-hexane at −35 °C. Single crystals were selected, coated with Parabar 10312 (previously known as Paratone N, Hampton Research, Aliso Viejo, CA, USA) and fixed on a microloop. Data were collected on a Bruker APEX DUO instrument (Billerica, MA, USA) equipped with an IµS microfocus sealed tube and QUAZAR optics for MoK_α radiation (λ = 0.71073 Å). The data collection strategy was determined using COSMO (Version 1.61, Bruker AXS Inc., Madison, WI, USA) employing ω- and φ scans. Raw data were processed using APEX 3 (Version 2016.5-0, Bruker AXS Inc., Madison, WI, USA) and SAINT (Version 8.37A, Bruker AXS Inc.,

Madison, WI, USA), corrections for absorption effects were applied using SADABS (Bruker AXS Inc., Madison, WI, USA, [51]). The structures were solved by direct methods and refined against all data by full-matrix least-squares methods on F^2 using SHELXTL [52,53] and SHELXLE [54]. Further details of the refinement and crystallographic data are listed in Table S4.1 (Supplementary Materials) and in the below mentioned CIF files. CCDCs 1817852 and 1817853 contain the supplementary crystallographic data for this paper. These data can be obtained free of charge from The Cambridge Crystallographic Data Centre via www.ccdc.cam.ac.uk/data_request/cif.

3.2. Synthesis of $Tb(AlMe_4)_3$ ($\mathbf{2}^{\mathbf{Tb}}$)

Homoleptic $\mathbf{2}^{\mathbf{Tb}}$ was synthesized according to the literature from $TbCl_3(thf)_{1.66}$ [4,11–13,20–23]. The chloride precursor was obtained via Soxhlet extraction of anhydrous $TbCl_3$ (5.00 g, 18.8 mmol) in THF at 100 °C for 7d, yielding 49% of the thf adduct. To a slurry of $TbCl_3(thf)_{1.66}$ (3.514 g, 9.128 mmol) in THF (40 mL) was added $LiNMe_2$ (1.397 g, 27.384 mmol) and the mixture was stirred for 1 d. Thereafter the solvent was removed under vacuum and 40 mL of *n*-hexane as well as $AlMe_3$ (5.915 g, 82.152 mmol) were added subsequently. After stirring for another day at ambient temperature, solid side products (e.g., LiCl) were removed by centrifugation and filtration and the solvent and by-product $(Me_2AlNMe_2)_2$ were removed under vacuum. Homoleptic $\mathbf{2}^{\mathbf{Tb}}$ was recrystallized three times from *n*-hexane (-35 °C, total crystallized yield 1.38 g, 3.29 mmol, 36%). ^{27}Al NMR (130 MHz, C_6D_6, 26 °C): $\delta = -203.3$ ppm. Elemental analysis calcd (%) for $C_{12}H_{36}Al_3Tb$ (420.28): C 34.29, H 8.63; found: C 33.88, H 9.11.

3.3. New Protocol for the Synthesis of $Lu(AlMe_4)_3$ ($\mathbf{2}^{\mathbf{Lu}}$)

In a 100 mL Schlenk flask, $LuCl_3(thf)_3$ (4.97 g, 10.0 mmol) was dissolved in THF (30 mL) and methyllithium (1.32 g, 60.0 mmol) was added under vigorous stirring. After 24 h, the solvent was removed under vacuum, the residue redissolved in toluene (10 mL) and the LiCl was removed via centrifugation and filtration. The toluene solution was evaporated in vacuo and the crude $[LuMe_3(MeLi)_3(thf)_x]$ was suspended in *n*-hexane (15 mL). Trimethylaluminum (8.64 g, 120.0 mmol) was added, the reaction mixture was stirred for 24 h and $LiAlMe_4$ was separated via filtration. Homoleptic $\mathbf{2}^{\mathbf{Lu}}$ crystallized from a saturated *n*-hexane solution at -35 °C (2.55 g, 5.84 mmol, 58%). ^{1}H NMR (400 MHz, C_6D_6, 26 °C): $\delta = -0.19$ (s, 36 H, $AlCH_3$) ppm. $^{13}C\{^1H\}$ NMR (100 MHz, C_6D_6, 26 °C): $\delta = 3.7$ ppm; elemental analysis calcd (%) for $C_{12}H_{36}Al_3Lu$ (436.33): C 33.03, H 8.32; found: C 33.08, H 8.61.

3.4. Polymerization of Isoprene

A detailed polymerization procedure is described as a typical example (Table S1.2.4, run 1). $[Ph_3C][B(C_6F_5)_4]$ (**A**) (18.2 mg, 0.02 mmol) was added to a solution of $\mathbf{2}^{\mathbf{Nd}}$ (8.1 mg, 0.02 mmol) in toluene (8 mL) and the mixture was aged at ambient temperature for 30 min. After the addition of isoprene (1.36 g, 20 mmol), the polymerization was carried out at 25 °C for 1 h. The reaction was terminated by pouring the polymerization mixture into 25 mL of methanol containing 0.1% (w/w) 2,6-di-tert-butyl-4-methylphenol as a stabilizer and stirred for 1 h. The polymer was washed with methanol and dried under vacuum at ambient temperature to constant weight.

Liquid cocatalysts (**D** and **E**) were added as 1 mL of a solution of a concentration of 1 equiv./mL. The amount of additional solvent applied was then reduced by 1 mL compared to the procedure given above in order to ensure highest possible similarity regarding the rheological properties of the polymerization mixture.

3.5. Polymerization of 1,3-Butadiene

A detailed polymerization procedure is described as a typical example (Table S1.6.3, run 1). $[Ph_3C][B(C_6F_5)_4]$ (**A**) (18.2 mg, 0.02 mmol) was added to a solution of $\mathbf{2}^{\mathbf{Pr}}$ (8.05 mg, 0.02 mmol) in toluene (28 mL) and the mixture was aged at ambient temperature for 30 min. The mixture was poured into an evacuated 50-mL Büchi miniclave and 1440 mL (60 mmol) of 1,3-butadiene were added at 90 mLn/min under constant stirring at 100 rpm. The polymerization was carried out at 25 °C for 1 h. After release of the remaining monomer pressure, if any, the reaction was terminated by pouring the polymerization mixture into 200 mL of methanol containing 0.1% (*w*/*w*) 2,6-di-tert-butyl-4-methylphenol as a stabilizer and stirred for 1 h. The polymer was washed with methanol and dried under vacuum at ambient temperature to constant weight.

Liquid cocatalysts (**D** and **E**) were added as 1 mL of a solution of a concentration of 1 equiv./mL. The amount of additional solvent applied was then reduced by 1 mL compared to the procedure given above in order to ensure highest possible similarity regarding the rheologic properties of the polymerization mixture.

3.6. Polymerization of Ethylene

A detailed polymerization procedure is described as a typical example (Table S1.8.1, run 1). Et_2AlCl (**E**) (3 mL of a solution of 0.02 mol/L in toluene, 1 equiv.) was added to a solution of $\mathbf{2^{Nd}}$ (24.3 mg, 0.06 mmol) in toluene (25 mL) and the mixture was aged at ambient temperature for 30 min. The mixture was poured into an evacuated 300-mL Parr autoclave, 100 mL toluene and 2100 mLn (60 mmol) of ethylene were added at a maximum flow rate of 450 mLn/min under constant stirring at 50 rpm. The polymerization was carried out at 60 °C for 1 h. After release of the remaining monomer pressure, if any, the reaction was terminated by pouring the polymerization mixture into 200 mL of methanol and stirred for 1 h. The polymer was washed with methanol and dried under vacuum at ambient temperature to constant weight.

3.7. Application of NMR Spectroscopic Techniques for Microstructure Determination

The microstructures of the polydiene samples produced in this study were evaluated by NMR spectroscopy Previously, the microstructure has been estimated measuring peak heights in the ^{13}C NMR spectra only [11,13]. Especially the vinylic content has been underestimated by this method as the variety in tacticity causes wide multiplets and therefore underrepresented peak heights. Furthermore, the nuclear Overhauser effect affects the vinylic and 1,4-polyisoprene signals significantly differently. The ratio of 1,4-/1,2 or 3,4-linkages is obtained from the ^{1}H NMR spectrum by integration of the respective multiplets in the vinylic region of the spectrum (approx. 4.5 to 4.9 ppm for the vinylic linkages and 4.9 to 5.5 ppm for the 1,4-linkages in case of polyisoprene and 4.5 to 5.15 ppm and 5.15 to 6.0 ppm, respectively, in case of polybutadiene). The *cis*/*trans* ratio of the 1,4-linked part of the polymer is obtained from its ^{13}C NMR spectrum, by integration of the respective methyl (methylene in case of 1,3-butadiene) peaks at 23.5 and 16 ppm in case of isoprene and 27.4 and 32 ppm, respectively, in case of 1,3-butadiene. In order to compare integration results in a proton decoupled ^{13}C NMR spectrum, the respective carbon atoms are assumed to be affected by relatively similar nuclear Overhauser effects, as they are decorated with the same number of hydrogen atoms and are located in similar chemical environments [55–58]. For meaningful evaluations, ^{13}C NMR spectra should show a low signal to noise ratio, requiring time-consuming data collection. Another drawback of this method is the fact that paramagnetic Ln ions, that have not been removed in the washing process, cause signal broadening via interaction with the polymer, which results in a loss of accuracy. Therefore, by far less time-consuming ATR-IR techniques have been probed in this study as a pre-screening method.

3.8. Application of ATR-Infrared Spectroscopic Techniques for Microstructure Detection

For pre-screening of the microstructure of the obtained polyisoprenes, ATR-IR spectroscopy was applied as a comparative method. NMR spectroscopic data were used for calibration. Hence, the obtained polymer data are considered to be less accurate than NMR-spectroscopy based data sets. But as ATR-IR spectroscopy is by far less costly and time-consuming, its suitability for pre-screening purposes was investigated. To ensure the lowest possible errors of the measurements, the comparative algorithm provided by the equipment manufacturer (Thermo Fisher Scientific, Waltham, MA, USA) was calibrated with a large data set of different polyisoprenes mirroring all possible microstructure compositions. In order to determine any deviation, both NMR and ATR-IR spectroscopic data were obtained and compared. Overviewing the findings (Figures S2.1.1–S2.4.1 and Tables S2.1.1–S2.4.1, Supplementary Materials) and considering the errors of the measurements, both methods are in good agreement. The highest individual absolute deviation found in the four series was 7.25%, on average 2.07%. Therefore, ATR-IR spectroscopy can very well serve as a pre-screening method for the microstructures of polyisoprenes. Furthermore, an optimized calibration of the algorithms might provide even more accurate results of wider applicability.

In order to provide an idea of the limits of the applicability of this method, surface manipulation of the polymers was performed by applying UV-light irradiation (Table S1.5.1, Supplementary Materials). Thereafter, the microstructure of the polymer samples was once more investigated by both techniques and compared. For NMR spectroscopy, no significant difference of the data before and after the manipulation was observed. But due to the low penetration depth of the IR-radiation into the polymer samples, in case of the ATR-IR data, such surface manipulation effects on the microstructure are emphasized (Table S2.5.1, Supplementary Materials). The observed shift toward vinylic linkage is supposed to be an artefact and due to an overlap of vibrations in the IR spectrum. UV-induced crosslinking is probably the origin of these new signals.

4. Conclusions

Homoleptic lanthanide and yttrium tris(tetramethylaluminate)s $Ln(AlMe_4)_3$ (Ln = La, Ce, Pr, Nd, Gd, Tb, Dy, Ho, Y, Er, and Lu) exhibit exceptional performance in 1,3-diene polymerization. Importantly, the reaction of $[LnMe_6(Li(do)_x)_3]$ with trimethylaluminum provides a more efficient precatalyst synthesis for the smaller rare-earth metal centers as shown for lutetium.

Due to the different nature of the borate/borane or R_2AlCl cocatalysts, distinct activation mechanisms occur, all providing individual trends and high-yielding precatalyst/cocatalyst combinations. The highest *cis*-1,4-contents were found when employing R_2AlCl cocatalysts and tetramethylaluminate complexes with Ln = Ce, Pr, Nd, and Gd, showing virtually complete *cis*-1,4-selectivity (>99.9% for Ln = Gd with cocatalyst Me_2AlCl). The highest *cis*-selectivity induced by the borate/borane cocatalysts was found for the combination of $Nd(AlMe_4)_3$ with 2 equiv. of $[B(C_6F_5)_3]$ (92.7%) although the general trend for these systems points to higher *cis*-selectivities with the smallest lanthanides. The most *trans*-selective combinations reach medium selectivity (<60%) for Ln = La and borate/borane cocatalysts. Interestingly, the uniformity of the active systems was best with Ln = Ce showing even a PDI < 1.2 with 2 equiv. of cocatalyst $[PhNMe_2H][B(C_6F_5)_4]$ and a surprisingly low PDI with Me_2AlCl, in comparison with the generally high PDI for the R_2AlCl-cocatalyzed polymers. Deeper insights into the polymerization performance of the binary system $Nd(AlMe_4)_3/Et_2AlCl$ could be gained by screening the pre-reaction time and the cocatalyst concentration. Here, different behaviors for the polymerization of 1,3-butadiene and isoprene were found. While the multimetallic cationic species, proposed in the literature seem to disaggregate upon monomer coordination for the bulkier isoprene, larger agglomerates are suggested to polymerize 1,3-butadiene. Interesting side reactions at high cocatalyst loadings as well as a comparatively high number of polymer chains per catalyst center, including further elucidation of the active species, will be subject to further research. Finally, ATR-IR spectroscopy displays an effective pre-screening method for the microstructures of polyisoprenes.

Supplementary Materials: The following are available online at www.mdpi.com/2073-4344/8/2/61/s1, Supplementary Materials (pdf) containing polymerization and crystal data. CIF files containing crystal data for complexes $\mathbf{2}^{\mathbf{Gd}}$ and $\mathbf{2}^{\mathbf{Tb}}$. The latter data have been deposited as well at CCDC (1817852 and 1817853).

Acknowledgments: We are grateful to the German Science Foundation (An238/14-2) and Bridgestone Japan for generous support, as well as Bassem Sabagh, Amy Ross (Malvern Instruments Ltd., Malvern, UK), and Bernd Schäfer (Malvern Instruments Ltd., Herrenberg, Germany) for performing the high temperature GPC measurements.

Author Contributions: C.O.H. and L.N.J. conceived and designed the experiments; D.D. contributed to the design of the ethylene polymerization experiments; C.O.H., L.N.J., D.D., T.Z. and R.S. performed the experiments; C.O.H., L.N.J., D.D., T.Z. and R.S. analyzed the data; C.M.-M. performed the X-ray structure analyses; C.O.H. wrote the paper; R.A. contributed reagents/materials/analysis tools, supervised all contributions and contributed to writing the paper.

Conflicts of Interest: The authors declare no conflict of interest.

References

1. Boor, J.J. *Ziegler-Natta Catalysts and Polymerizations*; Academic Press: New York, NY, USA; San Francisco, CA, USA; London, UK, 1979.
2. Mülhaupt, R. Catalytic Polymerization and Post Polymerization Catalysis Fifty Years After the Discovery of Ziegler's Catalysts. *Macromol. Chem. Phys.* **2003**, *204*, 289–327. [CrossRef]
3. Eisch, J.J. Fifty Years of Ziegler–Natta Polymerization: From Serendipity to Science. A Personal Account. *Organometallics* **2012**, *31*, 4917–4932. [CrossRef]
4. Fischbach, A.; Anwander, R. Rare-Earth Metals and Aluminum getting Close in Ziegler Type Organometallics. *Adv. Polym. Sci.* **2006**, *204*, 155–281.
5. Friebe, L.; Nuyken, O.; Obrecht, W. Neodymium-Based Ziegler/Natta Catalysts and their Application in Diene Polymerization. *Adv. Polym. Sci.* **2006**, *204*, 1–154.
6. Porri, L.; Ricci, G.; Giarrusso, A.; Shubin, N.; Lu, Z. Recent Developments in Lanthanide Catalysts for 1,3-Diene Polymerization. In *ACS Symposium Series 749—Olefin Polymerization: Emerging Frontiers*; Arjunan, P., McGrath, J.C., Hanlon, T., Eds.; Oxford University Press: New York, NY, USA, 2000; pp. 15–30.
7. Osakada, K.; Takeuchi, D. Coordination Polymerization of Dienes, Allenes, and Methylenecycloalkanes. In *Polymer Synthesis*; Springer: Berlin/Heidelberg, UK, 2004; pp. 137–194.
8. Shen, Z.; Ouyang, J. *Handbook on the Physics and Chemistry of Rare Earths*; Gschneidner, K.A., Jr., Fleming, L., Eds.; Elsevier Science Publishers: Amsterdam, The Netherlands, 1987; Chapter 61.
9. Taube, R.; Sylvester, G. *Applied Homogeneous Catalysis with Organometallic Compounds*; Cornils, B., Herrmann, W.A., Eds.; Wiley-VCH: Weinheim, Germany, 2002; pp. 280–318.
10. Zimmermann, M.; Anwander, R. Homoleptic Rare-Earth Metal Complexes Containing Ln−C σ-Bonds. *Chem. Rev.* **2010**, *110*, 6194–6259. [CrossRef] [PubMed]
11. Fischbach, A.; Klimpel, M.G.; Widenmeyer, M.; Herdtweck, E.; Scherer, W.; Anwander, R. Stereospecific Polymerization of Isoprene with Molecular and MCM-48-Grafted Lanthanide(III) Tetraalkylaluminates. *Angew. Chem. Int. Ed.* **2004**, *43*, 2234–2239. [CrossRef] [PubMed]
12. Zimmermann, M.; Frøystein, N.Å.; Fischbach, A.; Sirsch, P.; Dietrich, H.M.; Törnroos, K.W.; Herdtweck, E.; Anwander, R. Homoleptic Rare-Earth Metal(III) Tetramethylaluminates: Structural Chemistry, Reactivity, and Performance in Isoprene Polymerization. *Chem. Eur. J.* **2007**, *13*, 8784–8800. [CrossRef] [PubMed]
13. Meermann, C.; Törnroos, K.W.; Nerdal, W.; Anwander, R. Rare-Earth Metal Mixed Chloro/Methyl Compounds: Heterogeneous–Homogeneous Borderline Catalysts in 1,3-Diene Polymerization. *Angew. Chem. Int. Ed.* **2007**, *46*, 6508–6513. [CrossRef] [PubMed]
14. Fischbach, A.; Perdih, F.; Herdtweck, E.; Anwander, R. Structure−Reactivity Relationships in Rare-Earth Metal Carboxylate-Based Binary Ziegler-Type Catalysts. *Organometallics* **2006**, *25*, 1626–1642. [CrossRef]
15. Fischbach, A.; Perdih, F.; Sirsch, P.; Scherer, W.; Anwander, R. Rare-Earth Ziegler−Natta Catalysts: Carboxylate−Alkyl Interchange. *Organometallics* **2002**, *21*, 4569–4571. [CrossRef]
16. Arndt, S.; Beckerle, K.; Zeimentz, P.M.; Spaniol, T.P.; Okuda, J. Cationic Yttrium Methyl Complexes as Functional Models for Polymerization Catalysts of 1,3-Dienes. *Angew. Chem. Int. Ed.* **2005**, *44*, 7473–7477. [CrossRef] [PubMed]
17. Robert, D.; Spaniol, T.P.; Okuda, J. Neutral and Monocationic Half-Sandwich Methyl Rare-Earth Metal Complexes: Synthesis, Structure, and 1,3-Butadiene Polymerization Catalysis. *Eur. J. Inorg. Chem.* **2008**, 2801–2809. [CrossRef]
18. Litlabø, R.; Lee, H.S.; Niemeyer, M.; Törnroos, K.W.; Anwander, R. Rare-Earth metal bis(tetramethylaluminate) complexes supported by a sterically crowded triazenido ligand. *Dalton Trans.* **2010**, *39*, 6815–6825. [CrossRef] [PubMed]

19. Occhipinti, G.; Meermann, C.; Dietrich, H.M.; Litlabø, R.; Auras, F.; Törnroos, K.W.; Maichle-Mössmer, C.; Jensen, V.R.; Anwander, R. Synthesis and Stability of Homoleptic Metal(III) Tetramethylaluminates. *J. Am. Chem. Soc.* **2011**, *133*, 6323–6337. [CrossRef] [PubMed]
20. Nieland, A.; Mix, A.; Neumann, B.; Stammler, H.G.; Mitzel, N.W. Lanthanoid Tetramethylaluminates and Their Paramagnetic NMR Parameters. *Eur. J. Inorg. Chem.* **2014**, 51–57. [CrossRef]
21. Evans, W.J.; Anwander, R.; Ziller, J.W. Inclusion of Al_2Me_6 in the Crystalline Lattice of the Organometallic Complexes $LnAl_3Me_{12}$. *Organometallics* **1995**, *14*, 1107–1109. [CrossRef]
22. König, S.N.; Chilton, N.F.; Maichle-Mössmer, C.; Pineda, E.M.; Pugh, T.; Anwander, R.; Layfield, R.A. Fast magnetic relaxation in an octahedral dysprosium tetramethyl-aluminate complex. *Dalton Trans.* **2014**, *43*, 3035–3038. [CrossRef] [PubMed]
23. Klooster, W.T.; Lu, R.S.; Anwander, R.; Evans, W.J.; Koetzle, T.F.; Bau, R. Neutron Diffraction Study of $[Nd(AlMe_4)_3]$ $0.5\,Al_2Me_6$ at 100 K: The First Detailed Look at a Bridging Methyl Group with a Trigonal-Bipyramidal Carbon Atom. *Angew. Chem. Int. Ed.* **1998**, *37*, 1268–1270. [CrossRef]
24. Dietrich, H.M.; Raudaschl-Sieber, G.; Anwander, R. Trimethylyttrium and Trimethyllutetium. *Angew. Chem. Int. Ed.* **2005**, *44*, 5303–5306. [CrossRef] [PubMed]
25. Schumann, H.; Mueller, J.; Bruncks, N.; Lauke, H.; Pickardt, J.; Schwarz, H.; Eckart, K. Organometallic compounds of the lanthanides. Part 17. Tris[(tetramethylethylenediamine)lithium] hexamethyl derivatives of the rare earths. *Organometallics* **1984**, *3*, 69–74. [CrossRef]
26. Schumann, H.; Müller, J. Tris[(*N*,*N*,*N'*,*N'*-tetramethylethylenediamine)lithium]hexamethylerbate(III) and -lutetate(III). *Angew. Chem. Int. Ed.* **1978**, *17*, 276. [CrossRef]
27. Schumann, H.; Pickardt, J.; Bruncks, N. Crystal and Molecular Structure of$[Li(tmen)]_3[Er(CH_3)_6]$. *Angew. Chem. Int. Ed.* **1981**, *20*, 120–121. [CrossRef]
28. Shannon, R.D.; Prewitt, C.T. Effective Ionic Radii in Oxides and Fluorides. *Acta Cryst.* **1969**, *B25*, 925–946. [CrossRef]
29. Shannon, R.D. Revised Effective Ionic Radii and Systematic Studies of Interatomic Distances in Halides and Chalcogenides. *Acta Cryst.* **1976**, *A32*, 751–767. [CrossRef]
30. Chen, E.Y.-X.; Marks, T.J. Cocatalysts for Metal-Catalyzed Olefin Polymerization: Activators, Activation Processes, and Structure−Activity Relationships. *Chem. Rev.* **2000**, *100*, 1391–1434. [CrossRef] [PubMed]
31. Bonath, M.; Hollfelder, C.O.; Schädle, D.; Maichle-Mössmer, C.; Sirsch, P.; Anwander, R. C-H Bond Activation and Isoprene Polymerization by Lutetium Alkylaluminate/gallate Complexes Bearing a Peripheral Boryl and a Bulky Hydridotris(pyrazolyl)borato Ligand. *Eur. J. Inorg. Chem.* **2017**, 4683–4692. [CrossRef]
32. Diether, D.; Tyulyunov, K.; Maichle-Mössmer, C.; Anwander, R. Fluorenyl Half-Sandwich Bis(tetramethylaluminate) Complexes of the Rare-Earth Metals: Synthesis, Structure, and Isoprene Polymerization. *Organometallics* **2017**, *36*, 4649–4659. [CrossRef]
33. Monteil, V.; Spitz, R.; Boisson, C. Polymerization of butadiene and copolymerization of butadiene with styrene using neodymium amide catalysts. *Polym. Int.* **2004**, *53*, 576–581. [CrossRef]
34. Dietrich, H.M.; Schuster, O.; Törnroos, K.W.; Anwander, R. Heterobimetallic Half-Lanthanidocene Clusters: Novel Mixed Tetramethylaluminato/Chloro Coordination. *Angew. Chem. Int. Ed.* **2006**, *45*, 4858–4863. [CrossRef] [PubMed]
35. Evans, W.J.; Champagne, T.M.; Giarikos, D.G.; Ziller, J.W. Lanthanide Metallocene Reactivity with Dialkyl Aluminum Chlorides: Modeling Reactions Used to Generate Isoprene Polymerization Catalysts. *Organometallics* **2005**, *24*, 570–579. [CrossRef]
36. Evans, W.J.; Champagne, T.M.; Ziller, J.W. An Ethyl Aluminum Oxide (EAO) Complex with μ-η^1:η^2-Ethyl Coordination Derived from a Samarocene Carboxylate and Triethylaluminum. *Organometallics* **2005**, *24*, 4882–4885. [CrossRef]
37. Hayes, P.G.; Piers, W.E.; Parvez, M. Cationic Organoscandium β-Diketiminato Chemistry: Arene Exchange Kinetics in Solvent Separated Ion Pairs. *J. Am. Chem. Soc.* **2003**, *125*, 5622–5623. [CrossRef] [PubMed]
38. Yu, N.; Nishiura, M.; Li, X.; Xi, Z.; Hou, Z. Cationic Scandium Allyl Complexes Bearing Mono(cyclopentadienyl) Ligands: Synthesis, Novel Structural Variety, and Olefin Polymerization Catalysis. *Chem. Asian J.* **2008**, *3*, 1406–1414. [CrossRef] [PubMed]
39. Tardif, O.; Kaita, S. Generation of cationic indenyl silylamide gadolinium and scandium complexes $[(Ind)Ln\{N(SiMe_3)_2\}]^+[B(C_6F_5)_4]^-$ and their reactivity for 1,3-butadiene polymerization. *Dalton Trans.* **2008**, 2531–2533. [CrossRef] [PubMed]

40. Lei, Y.; Su, Q.; Chen, J.; Luo, Y. Synthesis, characterization of cationic half-sandwich scandium mono(silylamide) complexes and their unexpected reactivity toward C–Cl σ bond activation of chlorobenzene. *J. Organomet. Chem.* **2014**, *769*, 119–123. [CrossRef]
41. Zimmermann, M.; Törnroos, K.W.; Anwander, R. Cationic Rare-Earth-Metal Half-Sandwich Complexes for the Living *trans*-1,4-Isoprene Polymerization. *Angew. Chem. Int. Ed.* **2008**, *47*, 775–778. [CrossRef] [PubMed]
42. Tobisch, S.; Bögel, H.; Taube, R. Mechanistic Studies of the 1,4-Cis Polymerization of Butadiene According to the π-Allyl Insertion Mechanism. 1. Density Functional Study of the C−C Bond Formation Reaction in Cationic (η^3-Allyl)(η^2-/η^4-butadiene)nickel(II) Complexes $[Ni(C_3H_5)(C_4H6)]^+$ and $[Ni(C_3H_5)(C_4H_6)(C_2H_4)]^+$. *Organometallics* **1996**, *15*, 3563–3571.
43. Taube, R.; Windisch, H.; Maiwald, S. The catalysis of the stereospecific butadiene polymerization by Allyl Nickel and Allyl Lanthanide complexes—A mechanistic comparison. *Macromol. Symp.* **1995**, *89*, 393–409. [CrossRef]
44. Ward, B.D.; Bellemin-Laponnaz, S.; Gade, L.H. C_3 Chirality in Polymerization Catalysis: A Highly Active Dicationic Scandium(III) Catalyst for the Isoselective Polymerization of 1-Hexene. *Angew. Chem. Int. Ed.* **2005**, *44*, 1668–1671. [CrossRef] [PubMed]
45. Arndt, S.; Spaniol, T.P.; Okuda, J. Homogeneous Ethylene-Polymerization Catalysts Based on Alkyl Cations of the Rare-Earth Metals: Are Dicationic Mono(alkyl) Complexes the Active Species? *Angew. Chem. Int. Ed.* **2003**, *42*, 5075–5079. [CrossRef] [PubMed]
46. Dettenrieder, N.; Hollfelder, C.O.; Jende, L.N.; Maichle-Mössmer, C.; Anwander, R. Half-Sandwich Rare-Earth-Metal Alkylaluminate Complexes Bearing Peripheral Boryl Ligands. *Organometallics* **2014**, *33*, 1528–1531. [CrossRef]
47. Valente, A.; Mortreux, A.; Visseaux, M.; Zinck, P. Coordinative Chain Transfer Polymerization. *Chem. Rev.* **2013**, *113*, 3836–3857. [CrossRef] [PubMed]
48. Brintzinger, H.H.; Fischer, D.; Mülhaupt, R.; Rieger, B.; Waymouth, R.M. Stereospecific Olefin Polymerization with Chiral Metallocene Catalysts. *Angew. Chem. Int. Ed.* **1995**, *34*, 1143–1170. [CrossRef]
49. Rocha, T.C.J.; Coutinho, F.M.B.; Soares, B.G. Effect of alkylaluminum structure on Ziegler-Natta catalyst systems based on neodymium for producing high-cis polybutadiene. *Polym. Bull.* **2009**, *62*, 1–10. [CrossRef]
50. Hagihara, H.; Shiono, T.; Ikeda, T. Stereospecificity of propene polymerization with achiral titanocene-based catalysts. *Macromol. Chem. Phys.* **1998**, *199*, 2439–2444. [CrossRef]
51. Krause, L.; Herbst-Irmer, R.; Sheldrick, G.M.; Stalke, D. Comparison of silver and molybdenum microfocus X-ray sources for single-crystal structure determination. *J. Appl. Cryst.* **2015**, *48*, 3–10. [CrossRef] [PubMed]
52. Sheldrick, G.M. *SHELXT*—Integrated space-group and crystal-structure determination. *Acta Cryst.* **2015**, *A71*, 3–8. [CrossRef] [PubMed]
53. Sheldrick, G.M. Crystal structure refinement with *Shelxl*. *Acta Cryst.* **2015**, *C71*, 3–8.
54. Hubschle, C.B.; Sheldrick, G.M.; Dittrich, B. ShelXle: A Qt graphical user interface for SHELXL. *J. Appl. Crystallogr.* **2011**, *44*, 1281–1284. [CrossRef] [PubMed]
55. Elgert, K.-F.; Quack, G.; Stützel, B. Zur Struktur des Polybutadiens, 3. Das ^{13}C-NMR-Spektrum des *cis*-1,4-1,2-polybutadiens. *Makromol. Chem.* **1975**, *176*, 759–765. [CrossRef]
56. Elgert, K.-F.; Quack, G.; Stützel, B. Zur Struktur des Polybutadiens, 2. Das ^{13}C-NMR-Spektrum des 1,2-polybutadiens. *Makromol. Chem.* **1974**, *175*, 1955–1960. [CrossRef]
57. Elgert, K.-F.; Stützel, B.; Frenzel, P.; Cantow, H.-J.; Streck, R. Zur Struktur des Polybutadiens, 1. Das ^{13}C-NMR-Spektrum der 1,4-Polybutadiene. *Makromol. Chem.* **1973**, *170*, 257–260. [CrossRef]
58. Santee, E.R.; Chang, R.; Morton, M. 300 MHz proton NMR of polybutadiene: Measurement of *cis-trans* isomeric content. *J. Polym. Sci. Polym. Lett. Ed.* **1973**, *11*, 449–452. [CrossRef]

catalysts

MDPI

Article

Novel Allyl Cobalt Phosphine Complexes: Synthesis, Characterization and Behavior in the Polymerization of Allene and 1,3-Dienes

Giovanni Ricci [1,*], Antonella Caterina Boccia [1], Giuseppe Leone [1] and Alessandra Forni [2]

1 CNR, Istituto per lo Studio delle Macromolecole (ISMAC), via A. Corti 12, I-20133 Milano, Italy; antonella.boccia@ismac.cnr.it (A.C.B.); giuseppe.leone@ismac.cnr.it (G.L.)

2 CNR, Istituto di Scienze e Tecnologie Molecolari (ISTM), Università degli Studi di Milano, via C. Golgi 19, I-20133 Milano, Italy; alessandra.forni@istm.cnr.it

* Correspondence: giovanni.ricci@ismac.cnr.it; Tel.: +39-02-2369-9376

Received: 16 November 2017; Accepted: 1 December 2017; Published: 7 December 2017

Abstract: Novel allyl cobalt complexes, i.e., $(\eta^3\text{-}C_4H_7)(\eta^4\text{-}C_4H_6)Co(PCyPh_2)$ (**1**), $(\eta^3\text{-}C_4H_7)(\eta^4\text{-}C_4H_6)Co(PMePh_2)$ (**2**) and $(\eta^3\text{-}C_5H_9)(\eta^4\text{-}C_5H_8)Co(PMePh_2)$ (**3**), were synthesized by reacting $CoCl_2(PRPh_2)_2$ (R = methyl, cyclohexyl) with 1,3-butadiene or isoprene in presence of metallic zinc. The complexes were fully characterized by Nuclear Magnetic Resonance (NMR) spectroscopy (^{1}H and 2D experiments); in case of **1**, single crystals, suitable for X-ray analysis, were obtained and the molecular structure was determined. The allyl cobalt phosphine complexes alone gave highly crystalline 1,2 polymers from 1,2-propadiene, but they did not polymerize 1,3-dienes. Nevertheless, in the presence of a stoichiometric amount of methylaluminoxane (MAO), they were able to polymerize 1,3-butadiene and substituted 1,3-butadienes such as isoprene, (*E*)-1,3-pentadiene, (*E*)-1,3-hexadiene, and (*E*)-3-methyl-1,3-pentadiene. Specifically, **1**/MAO gave predominantly syndiotactic 1,2 polymers from 1,3-butadiene and terminally substituted 1,3-butadienes (e.g., 1,3-pentadiene and 1,3-hexadiene), but it was practically not active in the polymerization of internally substituted 1,3-butadienes (e.g., isoprene and 3-methyl-1,3-pentadiene); **2**/MAO and **3**/MAO exhibited instead an opposite behavior, giving predominantly isotactic 1,2 polymers from 3-methyl-1,3-pentadiene, and showing very low activity in the polymerization of 1,3-butadiene, 1,3-pentadiene and 1,3-hexadiene. The results obtained are interesting from the mechanistic point of view, and some hypotheses to explain this particular behavior were formulated.

Keywords: allyl cobalt phosphine complexes; X-ray structure; NMR characterization; 1,3-butadiene; polymerization of conjugated dienes; 1,2-polyallene

1. Introduction

Transition metal catalysts commonly used for the polymerization of 1,3-dienes can be grouped into the following classes: (i) catalysts derived from aluminum compounds (i.e., aluminum alkyls, aluminum hydrides and aluminoxanes) and transition metal compounds (Ziegler–Natta catalysts); (ii) catalysts derived from precursors not containing preformed metal-carbon bonds; (iii) catalysts based on allyl derivatives of transition metals [1–5]. Ziegler–Natta catalysts are obviously the most important ones, they can be highly active and stereospecific, and are industrially used for the production of *cis*-1,4 poly(1,3-butadiene), *cis*-1,4 polyisoprene and 1,2 syndiotactic poly(1,3-butadiene), the only diene polymers of industrial interest.

Extensive work has also been carried out with catalysts based on transition metal allyl derivatives [1]; these systems, which include simple allyl derivatives of transition metals and more complex catalysts derived from the reaction of the allyl complexes with electron donors, electron acceptors or Lewis acids, are of particular interest mainly because they are good models of the active site structure, and therefore provide useful information to elucidate the catalysis of polymerization.

The first examples of stereospecific polymerization induced by simple allyl derivatives of transition metals were first reported in 1964. A predominantly 1,2 poly(1,3-butadiene) was obtained with (allyl)$_3$Cr and a predominantly *cis*-1,4 polymer with (allyl)$_2$CoI [6]. A crystalline poly(1,3-butadiene) consisting almost exclusively of *trans*-1,4 units and a predominantly *cis*-1,4 polymer were obtained with (allyl)NiBr depending on the type of polymerization solvent used [7]. Allyl derivatives of other transition metals (Ti, Zr, Mo, W, Ru, Rh, U, Nd) were later found to be effective catalysts [8–10]. More recently, the allyl cobalt compound (η^4-C_4H_6)(η^5-C_8H_{13})Co, alone or in presence of CS_2, was reported to be highly active and stereoselective in the polymerization of 1,3-butadiene to 1,2 syndiotactic polymer [11–15].

We have now synthesized three novel allyl cobalt compounds (η^3-C_4H_7)(η^4-C_4H_6)Co(PCyPh$_2$) (**1**), (η^3-C_4H_7)(η^4-C_4H_6)Co(PMePh$_2$) (**2**) and (η^3-C_5H_9)(η^4-C_5H_8)Co(PMePh$_2$) (**3**). The complexes were characterized by Nuclear Magnetic Resonanance (NMR) (^{1}H and 2D experiments) and X-ray diffraction analysis, and their behavior in the polymerization of 1,2-propadiene and various 1,3-dienes was investigated. The most significant results obtained are reported in this paper.

2. Results

2.1. *Synthesis and Characterization of the Cobalt Phosphine Complexes*

The allyl cobalt phosphine **1–3** complexes were prepared by reacting CoCl$_2$(PRPh$_2$)$_2$ (R = methyl, cyclohexyl) with 1,3-butadiene or isoprene in presence of metallic zinc (Scheme 1), according to the experimental procedure already described in the literature for analogous complexes [16,17].

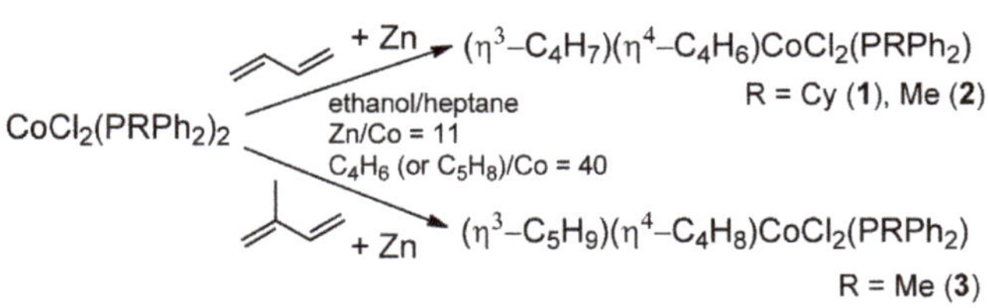

Scheme 1. Synthesis of the allyl cobalt phosphine complexes.

2.1.1. X-ray Crystal Structure and Computational Analysis

In case of (η^3-C_4H_7)(η^4-C_4H_6)Co(PCyPh$_2$) (**1**), single crystals suitable for X-ray structure determination were obtained. The molecular structure is shown in Figure 1, while selected bond lengths and angles, and crystallographic data are reported in Tables 1 and S1, respectively. Structural analysis reveals that the coordination sphere around the cobalt ion comprises, besides the phosphine ligand, the C1–C3 allyl group and the C5–C8 butadiene molecule, with the C1, C2 allylic carbons and the C6, C7 central atoms of butadiene closest to Co with respect to C3 and the butadiene lateral carbons C5 and C8, respectively.

The allylic unit and the butadiene molecule face approximately parallel to each other, the dihedral angle between their respective least-squares planes measuring 13(1)°. Within the phosphine unit, the P–C bond lengths are slightly shorter for the aromatic with respect to the aliphatic carbon atoms, as expected. Comparison with the only structurally related cobalt complex reported in literature, i.e., (η^3-C_4H_7)(η^4-C_4H_6)Co(PPh$_3$) [16], shows that the most evident differences with the present structure are found in the torsion angle C1–C2–C3–C4, θ, which is 47(2)° in the former and −173.9(7)° in the latter, and in the spatial disposition of the phosphinic ligand with respect to the allyl/butadiene moieties. Such conformational differences appear to be correlated. In fact, considering the plane containing the Co–P bond and roughly parallel to the C1–C3 and C5–C8 least-squares planes (which are not strictly parallel as noted above), one phenyl ring of the structure with PPh_3 [16] lies approximately in this plane, i.e., midway between the allylic unit and the butadiene molecule, and points on the same side as the C4 methyl group.

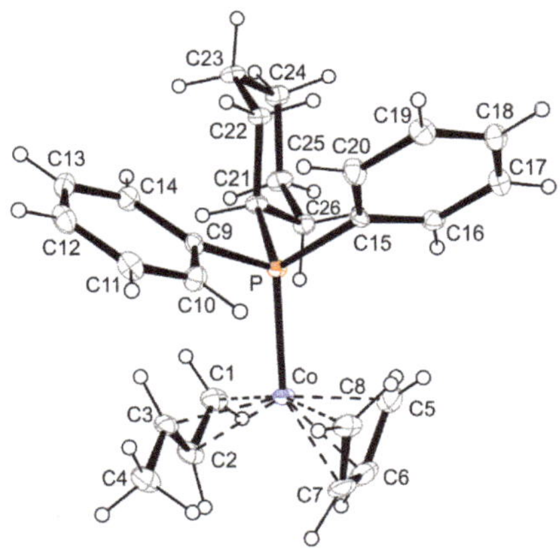

Figure 1. Molecular structure of (η^3-C_4H_7)(η^4-C_4H_6)Co($PCyPh_2$) (**1**) with thermal ellipsoids drawn at 30% probability level.

The consequent Ph/Me repulsion is alleviated by a moderate deviation of the methyl from the perfect, more stable, *anti* geometry ($\theta = 0°$). In the present structure, on the other hand, the $PCyPh_2$ ligand is rotated around the Co–P bond in such a way as to bring the bulkier Cy ring as far away from the methyl group. The large Ph/Me repulsion which would arise in such a conformation is avoided by a further rotation of the methyl towards the less stable *syn* geometry ($\theta = 180°$).

Table 1. Selected X-ray Bond Lengths (Å) and Angles (°) for complex **1**.

Co–P	2.2049(16)	C1–C2	1.406(10)	C1–Co–P	99.3(2)
		C2–C3	1.405(10)	C2–Co–P	120.3(2)
Co–C1	2.066(7)	C3–C4	1.503(10)	C3–Co–P	96.8(2)
Co–C2	1.996(7)	C5–C6	1.405(12)	C5–Co–P	99.5(2)
Co–C3	2.123(7)	C6–C7	1.420(12)	C6–Co–P	137.7(3)
Co–C4	3.192(8)	C7–C8	1.397(10)	C7–Co–P	134.3(2)
Co–C5	2.083(8)			C8–Co–P	95.3(2)
Co–C6	2.010(7)	P–C9	1.841(6)		
Co–C7	2.006(7)	P–C15	1.839(6)		
Co–C8	2.085(7)	P–C21	1.854(6)		

Subtler but not negligible differences are found in bond lengths. The Co–P bond is elongated by 0.014 Å (from 2.191(4) [16] to 2.2049(16) Å) as a result of the replacement of one aromatic phenyl unit with the aliphatic cyclohexyl ring, which decreases the π–character of the Co–P bond and increases the steric hindrance of the phosphine ligand. On the side of the π–system coordination, on the contrary, the Co–C bond lengths are systematically shorter in the present structure, suggesting a greater π–character for these bonds. Although such differences in bond lengths are at the limit of the experimental accuracy, they are partially reproduced by Density Functional Theory (DFT) calculations in gas phase (Table 2). They provide a Co–P lengthening, equal to 0.021 Å, and a slight shortening in the Co–C(allyl) bond lengths going from PPh_3 to $PCyPh_2$. The butadiene molecule is instead approximately at the same distance from the Co ion. Interestingly, the simulated structure of (η^3-C_4H_7)(η^4-C_4H_6)Co($PMePh_2$) (**2**) features the shorter Co–P and Co–C(allyl) bonds within the series of the three complexes. The $PMePh_2$ phosphine ligand shares in fact the same π–character as $PCyPh_2$, but its sterical hindrance is drastically lower, allowing a significant approaching to the cobalt ion.

Table 2. Selected M06/6-311g(d) computed bond lengths (Å) for (η^3-C_4H_7)(η^4-C_4H_6)Co($PRPh_2$) with R = cyclohexyl (Cy), phenyl (Ph) and methyl (Me).

Bond	$PCyPh_2$	PPh_3	$PMePh_2$
Co–P	2.233	2.212	2.206
Co–C1	2.044	2.050	2.039
Co–C2	1.995	1.999	1.990
Co–C3	2.089	2.093	2.084
Co–C5	2.053	2.053	2.045
Co–C6	1.999	2.001	1.999
Co–C7	2.006	2.003	2.006
Co–C8	2.058	2.049	2.056
C1–C2	1.412	1.409	1.413
C2–C3	1.403	1.403	1.406
C3–C4	1.495	1.496	1.496
C5–C6	1.410	1.411	1.411
C6–C7	1.415	1.415	1.415
C7–C8	1.411	1.413	1.411

In the crystal structure of complex **1**, several H–H intramolecular distances are found below or slightly above two times the H atom van der Waals radius, involving the butadiene and the allyl group (H2–H6, 2.41 Å; H4C–H7, 2.23 Å); the butadiene and either phenyl groups (H5A–H20, 2.25 Å; H8A–H10, 2.41 Å), or the Cy ring (H5B–H26B, 2.49 Å); the allylic group and the Cy ring (H1B–H21, 2.50 Å and H1B–H26B, 2.52 Å); the Ph and the Cy phosphinic units (H14–H21, 2.18 Å, H14–H22A, 2.16 Å and H20–H26A, 2.33 Å).

2.1.2. NMR Characterization

The ^{1}H NMR spectra of (η^3-C_4H_7)(η^4-C_4H_6)Co($PCyPh_2$) (**1**) and (η^3-C_4H_7)(η^4-C_4H_6)Co($PMePh_2$) (**2**) are shown in Figure 2a,b, respectively. Protons of the allyl unit and of the coordinated butadiene monomer were assigned referring to the integrals, multiplicity, and chemical shifts in the ^{1}H NMR spectra, and to the ^{1}H–^{1}H correlations in TOtal Correlation SpectroscopY (TOCSY) experiments (Figures 3 and S1). Information concerning the spatial reciprocal orientation of the allylic unit and of the butadiene monomer coordinated to the cobalt atom can be deduced from the ^{1}H–^{1}H correlations in Rotating frame Overhauser Effect SpectroscopY (ROESY) experiments (see below in the text).

The ^{1}H spectrum of **1** (Figure 2a) was assigned as follows: H_i and H_e, multiplet, as sum of two triplets, at $\delta = -0.24$ ppm; H_d, triplet at $\delta = 0.30$ ppm, J_{db} 11 Hz; Me, doublet at $\delta = 0.87$ ppm; H_c, merging signal at $\delta = 1.02$ ppm; H_k and H_f, two doublets at $\delta = 1.60$ and $\delta = 2.22$ ppm; H_a, broad doublet at $\delta = 1.77$ ppm; H_b, multiplet at $\delta = 4.26$ ppm; H_g and $H_{g'}$, two multiplets at $\delta = 4.36$ and 4.45 ppm respectively, J_{gk} 7 Hz, J_{ge} 12 Hz; H_{cy}, signals at $\delta = 0.80, 0.99, 1.21, 1.59, 2.22$, and 2.42 ppm; H_{ar}, doublet at $\delta = 7.13$ ppm, and two overlapping triplets at $\delta = 7.62$ and 7.68 ppm, respectively. The up-field shift for the anti (H_e and H_i) protons of butadiene, Figure 2a, seems to suggest a partial σ character of the bond between the cobalt atom and C1, C4 carbon atoms of coordinated butadiene, in agreement with those previously reported [17]. Protons H_g and $H_{g'}$ are low field shifted as expected for the olefin carbons C2 and C3 of the butadiene unit. To distinguish the $H_g/H_{g'}$ protons from H_b and to confirm the above reported assignment the ^{1}H–^{1}H TOCSY experiments (Figure 3) were of a great help. Because only $H_{i,e}$ and $H_{g,g'}$ are relative to the same spin system, it is expected that they generate two partially merging cross peaks on the TOCSY spectrum (namely 1 and 2 in Figure 3). Moreover, the protons of the cyclohexyl group of the phosphine ligand were unambiguously solved also if present in a very crowded region in the ^{1}H spectrum.

The ^{1}H NMR spectrum of complex **2** (Figure 2b) was assigned as follows, by applying the same strategy reported above for the other cobalt complex: $H_{i,e}$ protons, two triplets at $\delta = -1.21$ ppm and -0.51 ppm, $J_{i,e/g,g'}$ 11 Hz; methyl protons of the allylic unit, two superimposed doublets at $\delta =$ 0.37 ppm, $J_{Me/a,b}$ 7 Hz; H_d, triplet at $\delta = 0.93$ ppm; methyl protons of the methyl group bonded to the phosphorus atom, doublet at $\delta = 1.96$ ppm, $J_{Me,d}$ 6.7 Hz; $H_{k,f}$, two doublets at $\delta = 2.08$ and 2.40 ppm,

$J_{k,f/g,g'}$ 7 Hz; H_c, merging signal with $H_{k,f}$ at δ = 2.08 ppm; H_a, multiplet at δ = 2.79 ppm; $H_{g,g'}$, two multiplets at δ = 4.23 and 4.47 ppm; H_b, multiplet at δ = 4.58 ppm; aryl protons H_{ar} of the phosphine ligand at δ = 7.09, 7.55, and 7.68 ppm. Uncertainties about some closer resonances were solved by TOCSY experiments (Figure S1). In fact, H_g and $H_{g'}$ were distinguished from H_b as they are part of the same spin system and they have to correlate with each other.

complex\ppm	H_a	H_b	H_c	H_d	H_e/H_i	H_f/H_k	$H_g/H_{g'}$	Me
1 *syn* (η^3-C_4H_7)(η^4-C_4H_6)Co($PCyPh_2$)	1.77	4.26	1.02	0.30	-0.24	1.60 2.22	4.36 4.45	0.87
2 *anti* (η^3-C_4H_7)(η^4-C_4H_6)Co($PMePh_2$)	2.79	4.58	2.08	0.93	-1.21 -0.51	2.08 2.40	4.23 4.47	0.37

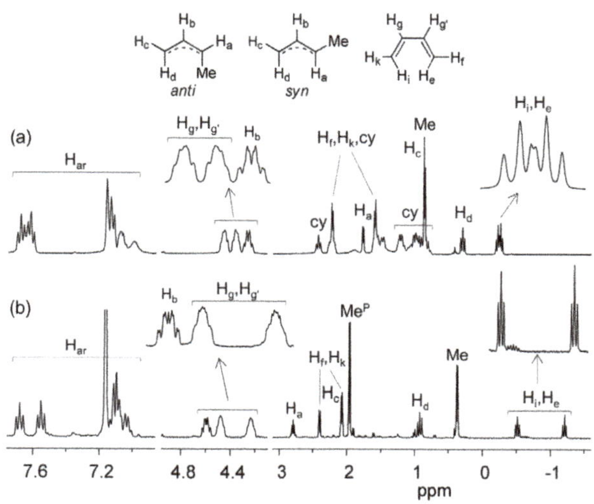

Figure 2. ^{1}H NMR spectra of (**a**) (η^3-C_4H_7)(η^4-C_4H_6)Co($PCyPh_2$) (**1**) and (**b**) (η^3-C_4H_7)(η^4-C_4H_6)Co($PMePh_2$) (**2**) (Me^P indicates the protons of the methyl on the ligand phosphorus atom, H_{ar} = aryl protons, cy = cyclohexyl protons).

The ^{1}H NMR spectrum of (η^3-C_5H_9)(η^4-C_5H_8)Co($PMePh_2$) (**3**) is shown in Figure 4, and it was assigned as follows. ^{1}H spectrum: H_e, doublet at δ = −1.40 ppm, $J_{e/f}$ 13 Hz; H_i, triplet at δ = −0.56 ppm, $J_{i/g}$ 11 Hz; H_{Me}, singlet at δ = 0.38 ppm; H_d, triplet at δ = 0.67 ppm, $J_{d/b}$ 12 Hz; H_{Me}, singlet at δ = and 1.21 ppm; H_c, doublet at δ = 1.55 ppm, $J_{c/b}$ 7 Hz; $H_{Me''}$, doublet at δ = 1.62 ppm; H_f, singlet at δ = 1.78 ppm; H_k, doublet at δ = 1.93 ppm; $H_{Me/P}$, signal partially merging with H_k at δ = 1.93 ppm; H_b, broad triplet at δ = 4.20 ppm; H_g, triplet at δ = 4.35 ppm, $J_{g/k}$ 8 Hz; H_{ar}, signals at δ = 7.13, 7.50 and 7.68 ppm. ^{1}H–^{1}H TOCSY experiment confirms the exact assignment of protons spin system (Figure S2).

On the basis of the above, it appears evident that the most significant ^{1}H NMR signals permitting to discriminate between the three different complexes are those related to the following protons: allyl methyl (Me), $H_{i/e}$, $H_{g/g'}$ and H_b. The ppm values of the allyl methyl protons are clearly indicative of the type of allyl unit configuration (*anti* or *syn*) [17], while the ppm values and multiplicities of $H_{i/e}$, $H_{g/g'}$ and H_b are most likely related to the type of reciprocal monomer/allyl unit orientation (*exo-exo* or *exo-endo*). Further work is certainly needed (e.g., synthesis of allyl cobalt complexes with different phosphine ligands; further NMR experiments, also at variable temperature) in order to confirm these indications.

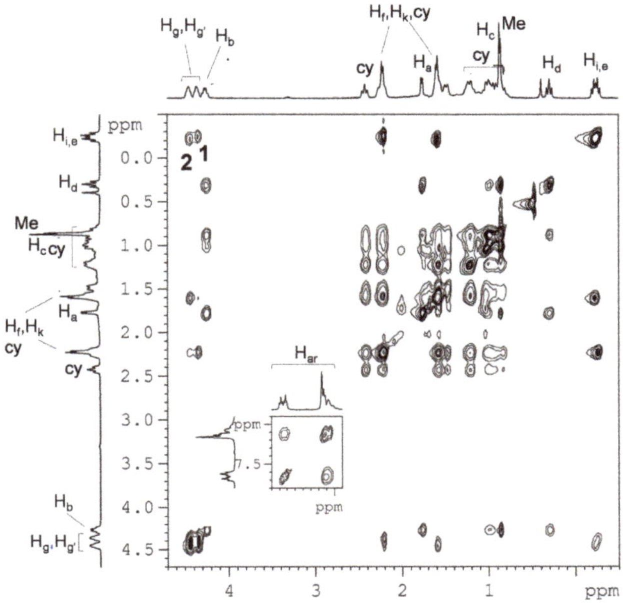

Figure 3. TOCSY NMR spectrum of $(\eta^3\text{-}C_4H_7)(\eta^4\text{-}C_4H_6)Co(PCyPh_2)$ (**1**).

complex\ppm	H_a	H_b	H_c	H_d	H_e	H_i	H_f	H_g	H_k	Me_e	$Me_{e'}$	$Me_{e''}$	Me^P
3 *syn* $(\eta^3\text{-}C_5H_9)(\eta^4\text{-}C_5H_8)$ $Co(PMePh_2)$	-	4.20	1.55	0.67	-1.40	-0.56	1.78	4.35	1.93	0.38	1.21	1.62	1.93

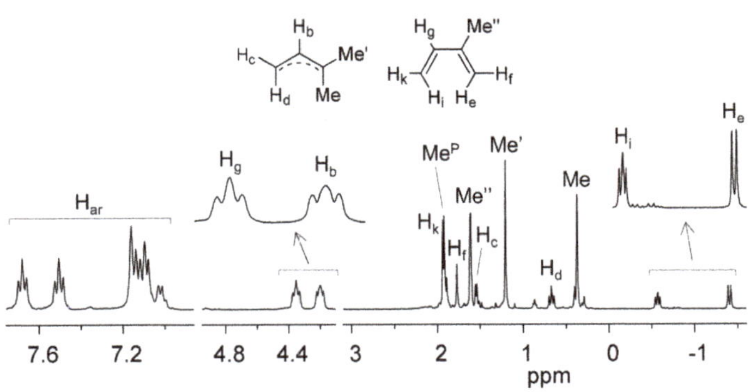

Figure 4. ^{1}H NMR spectra of $(\eta^3\text{-}C_5H_9)(\eta^4\text{-}C_5H_8)Co(PMePh_2)$ (**3**) (Me^P indicates the protons of the methyl on the ligand phosphorus atom, H_{ar} = aryl protons).

In order to get information about the reciprocal spatial rearrangement of the allylic unit and of the butadiene (or isoprene) monomer in the three different cobalt complexes synthesized, ^{1}H–^{1}H ROESY experiments were acquired (Figures 5, S3 and S4). In case of ^{1}H–^{1}H ROESY experiment of complex **1** (Figure 5), correlations of the aromatic aryl protons on the phosphorus atom, at δ = 7.68 ppm, with some aliphatic protons, put in evidence the spatial vicinity between the phosphine ligand and the allylic and butadiene units, thus providing information on the reciprocal orientation of these two units. As can be observed in Figure 5, aryl protons of the phosphine ligand at 7.68 ppm correlate with (i) the cyclohexyl protons of the same phosphine ligand; (ii) the protons H_i and H_e of the butadiene monomer (*cross-peak 1*); (iii) the H_d and H_a protons of the allylic unit (*cross-peaks 2* and *3*, respectively). Moreover, the cyclohexyl proton at $\delta = 2.42$ ppm correlates with the $H_{i,e}$ and H_d protons (*cross-peaks 4* and *5*) of the allylic and butadiene unit, respectively, while no correlations are observed with the protons $H_{g,g'}$ and H_b, thus in agreement with the reciprocal *exo-exo* spatial orientation of the allylic unit and the butadiene monomer shown by the X-ray molecular structure of complex **1** (Figure 1). Finally, a *syn* configuration for the allyl unit was suggested by the strong correlation between H_a and H_d protons, again in agreement with the crystalline structure of **1**.

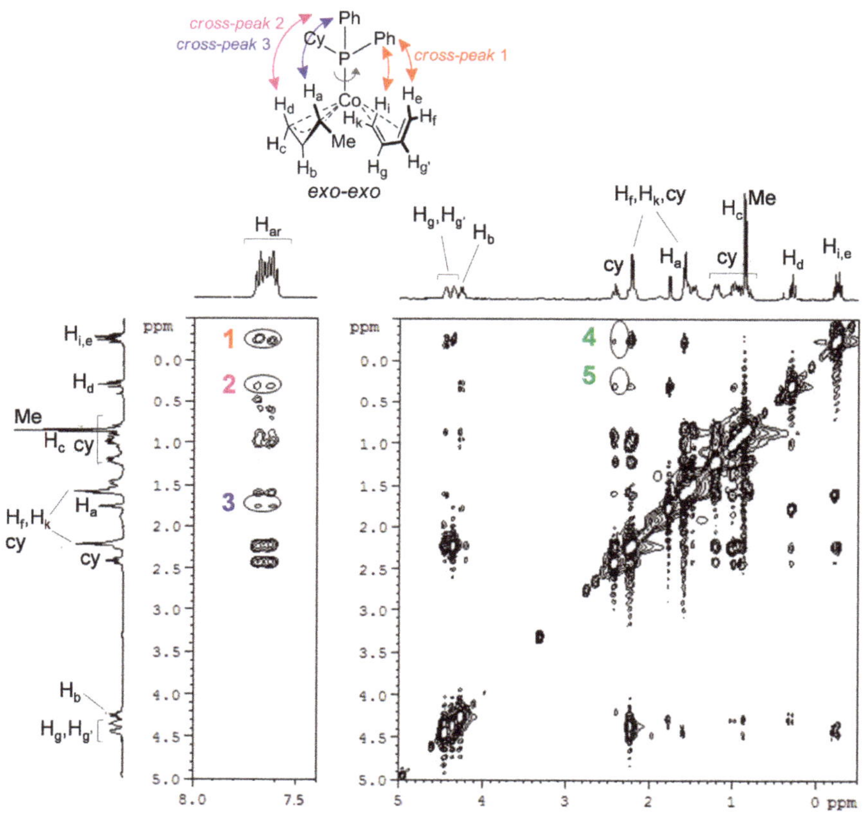

Figure 5. ROESY NMR spectrum of $(\eta^3\text{-}C_4H_7)(\eta^4\text{-}C_4H_6)Co(PCyPh_2)$ (**1**).

As concern the reciprocal spatial orientation of the isoprene monomer and of allylic unit in case of $(\eta^3\text{-}C_5H_9)(\eta^4\text{-}C_5H_8)Co(PMePh_2)$ (**3**), the ROESY experiments (Figure S3) seem instead to suggest the presence of an *exo-endo* orientation, or anyway the coexistence of both the orientations, *exo-endo* (Figure 6A) and *exo-exo* (Figure 6B). The methyl-*anti* (Me) at 0.38 ppm and methyl-*syn* (Me′) protons at 1.21 ppm of the allylic unit spatially correlate with the H-aryl's on the phosphine (*cross peaks 1* and *2*, the *cross-peak 2* being less intense because "*less close*" to the corresponding Me′ group (Figure S3)). In addition, as only Me protons at $\delta = 0.38$ ppm correlate with Me on the phosphine (Me^P at 1.93 ppm), it is possible to affirm that Me-*anti* is oriented on the same side of the phosphine ligands, confirming indeed the *syn/anti* attribution of the two methyl groups Me and Me′ and the *exo* orientation of the allylic unit. Moreover, the H_g and $H_{Me'}$ interaction (*cross peak 3*, Figure S3) seems to suggest that they are oriented on the same side, that is compatible with an *exo-endo* reciprocal orientation of the allylic group and of the isoprene unit (Figure 6A); on the other hand the complete absence of any interactions between $H_{i,e}$ and $H_{f,k}$ with the aryl and methyl protons of the phosphine ligand are also in favor of an *exo-endo* orientation of the allylic group and of the isoprene unit. This means that when the phosphine ligand is $PMePh_2$, both the arrangements (*exo-exo* and *exo-endo*) are possible, in agreement indeed with the formation of atactic 1,2 polymers in the polymerization of terminally substituted monomers such as 1,3-pentadiene and 1,3-hexadiene [18] with catalyst based on complex **3**, as reported below in the text.

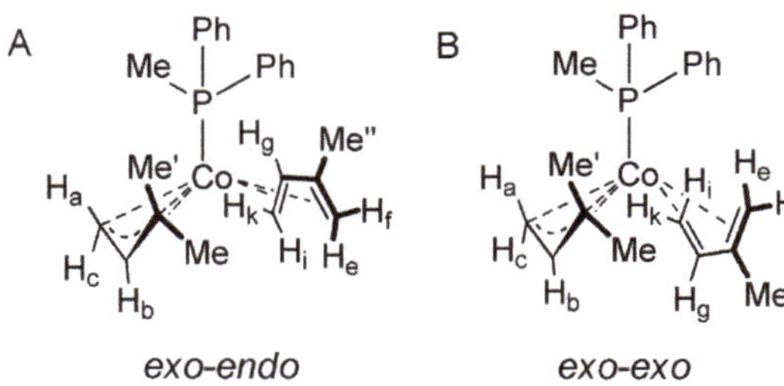

Figure 6. Possible reciprocal spatial orientations of the allyl group and the isoprene monomer for $(\eta^3\text{-}C_5H_9)(\eta^4\text{-}C_5H_8)Co(PMePh_2)$ (**3**): *exo-endo* (**A**) and *exo-exo* (**B**).

The ROESY experiments carried out on complex **2** (Figure S4) seem to suggest a situation very close to that found for complex **3**. No correlations involving allylic protons (H_a, H_b, H_c) and the protons of the aryl phosphine ligand were detected, while some experiments reveal a weak correlation between the methyl group (*syn*) of the allylic unit and the aryl protons, that is what expected for an *exo* rearrangement of the allylic group. As regard the orientation of the 1,3-butadiene monomer, in all the ROESY experiments carried out, only one proton of the butadiene monomer, H_f or H_k, at δ = 2.40 ppm, (*cross-peak 1*, Figure S4), was found to spatially correlate with protons of the aryl phosphine ligand, at δ = 7.68 and 7.55 ppm, suggesting indeed an *exo* orientation of butadiene; however, the H_g proton was found to correlate with the methyl protons of the allylic group, in agreement in this case with an *endo* orientation of the butadiene monomer. It follows indeed that, as occurred in the case of complex **3**, both orientations, *exo-exo* and *exo-endo*, are possible, indicating some fluxionality of the butadiene monomer coordinated to the cobalt atom.

2.2. Polymerization

2.2.1. Polymerization of 1,2 Propadiene

The allyl cobalt complexes alone do not polymerize 1,3-dienes, as we will see later, while they show a rather low activity in the polymerization of 1,2-propadiene (Table 3), giving highly stereoregular crystalline 1,2 polymers with a melting point of about 110 °C (Figure 7a,b). As an example, the ^{13}C NMR spectrum of the poly(1,2-propadiene) obtained with $(\eta^3\text{-}C_4H_7)(\eta^4\text{-}C_4H_6)Co(PMePh_2)$ (Table 3, run 2) is shown in Figure 7c. It is worth noting that complex **3** was more active than **1** and **2**, likely due to its lower stability and/or greater reactivity toward the interaction with MAO.

Table 3. Polymerization of 1,2-propadiene with cobalt complexes **1–3** [a].

Entry	Catalyst	Time (h)	Yield (g)	M_w ($\times 10^3$)	M_w/M_n
1	1/MAO	2	0.05	22	3.2
2	2/MAO	1.5	0.17	21	3.8
3	2	21	0.14	18	2.8
4	3/MAO	1	0.59	35	2.9

[a] Polymerization conditions: 1,2 propadiene, 2 mL; toluene, total volume 16 mL; MAO/Co = 2; Co, 5×10^{-5} mol; temperature, 22 °C.

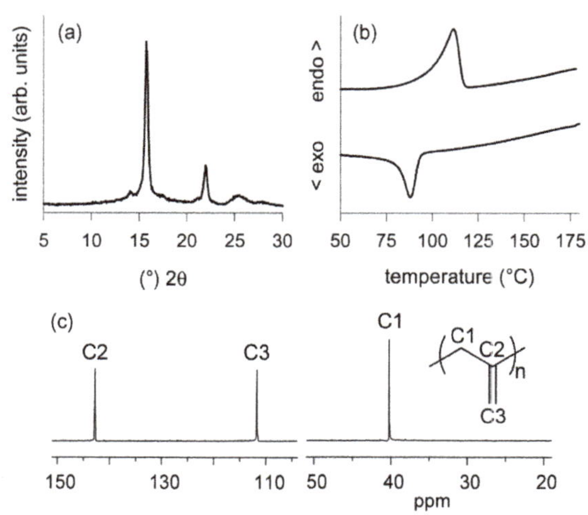

Figure 7. Powder X-ray Diffraction (PXRD) (**a**); Differential Scanning Calorimetry (DSC) (**b**) and ^{13}C NMR (**c**) spectra of poly(1,2-propadiene) (Table 3, entry 2).

2.2.2. Polymerization of 1,3 Dienes

As mentioned above, the allyl cobalt complexes alone do not polymerize 1,3-dienes. They become, however, active in the presence of small amounts (stoichiometric) of MAO, as shown in Table 4. Predominantly 1,2 polymers were obtained from 1,3-butadiene; (η^3-C_4H_7)(η^4-C_4H_6)Co($PCyPh_2$) (**1**) exhibited a quite good activity, giving predominantly syndiotactic polymers, while (η^3-C_4H_7)(η^4-C_4H_6)Co($PMePh_2$) (**2**) and (η^3-C_5H_9)(η^4-C_5H_8)Co($PMePh_2$) (**3**) were characterized by a rather low activity and stereospecificity, giving essentially atactic 1,2 polymers.

The allyl cobalt complexes were also able to polymerize substituted 1,3-butadienes such as (*E*)-1,3-pentadiene, (*E*)-1,3-hexadiene, isoprene and (*E*)-3-methyl-1,3-pentadiene. Complex **1** resulted to be highly active and selective in the polymerization of 1,3-pentadiene and 1,3-hexadiene, giving crystalline, highly syndiotactic 1,2 polymers, while it was almost inactive in the polymerization of isoprene and 3-methyl-1,3-pentadiene.

On the contrary, complexes **2** and **3** were highly active in the polymerization of isoprene and 3-methyl-1,3-pentadiene, while only polymer traces were obtained from 1,3-pentadiene and 1,3-hexadiene. Furthermore, they gave a predominantly isotactic polymer from 3-methyl-1,3-pentadiene and an alternate *cis*-1,4-*alt*-3,4 polymer from isoprene [19].

Table 4. Polymerization of 1,3-dienes with cobalt complexes **1–3** [a].

Entry	Catalyst	Time (h)	Yield (g)	Conv. (%)	1,2 [b] (mol%)	rr/mr/mm [c]	M_w ($\times 10^3$)	M_w/M_n
1,3-butadiene								
1	1	168	-					
2	2	168	-					
3	3	168	-					
4	1/MAO	2	0.87	62	87	71/25/4	169	2.4
5	2/MAO	15	0.20	14	73	26/48/26	160	2.5
6	3/MAO	10	traces					
E-1,3-pentadiene								
7	1/MAO	15	1.24	91	≥99	69/31/0	150	1.7
8	2/MAO	24	0.26	19	≥99	24/50/26	97	1.6
1,3-hexadiene								
9	1/MAO	12	0.66	46	≥99	72/28/0	74	1.9
10	2/MAO	144	0.18	13	≥99	22/51/27	55	1.6
isoprene								
11	1/MAO	24	traces					
12	2/MAO	24	0.45	33	~50		68	1.5
13	3/MAO	24	0.51	38	~50		75	1.4
3-methyl-1,3-pentadiene								
14	1/MAO	24	traces					
15	2/MAO	24	0.94	64	100	0/0/100	90	1.2
16	3//MAO	24	0.85	58	100	0/0/100	87	1.3

[a] Polymerization conditions: monomer, 2 mL; toluene as solvent, total volume 16 mL; MAO/Co molar ratio 2–3; Co, 5×10^{-5} mol; temperature, 22 °C. [b] mol% of 1,2 units (3,4 in case of isoprene), the remaining units are essentially *cis*-1,4. [c] percentage of syndiotactic/atactic/isotactic triads, determined by ^{13}C NMR.

3. Discussion

The reasons why (i) the above allyl complexes become active only when used in combination with MAO and (ii) some complexes (e.g., **1**) are highly active and selective in the polymerization of 1,3-butadiene, 1,3-pentadiene and 1,3 hexadiene (i.e., terminally substituted 1,3-butadienes) and not in case of isoprene and 3-methyl-1,3-pentadiene (i.e., internal substituted 1,3-butadienes), while other complexes (e.g., **2** and **3**) exhibit a completely opposite behavior being highly active and selective in the polymerization of isoprene and 3-methyl-1,3-pentadiene (i.e., internal substituted 1,3-butadienes) and not in case of 1,3-butadiene, 1,3-pentadiene and 1,3-hexadiene, are somewhat difficult to explain and to rationalize.

A tentative interpretation for the role of MAO is shown in Scheme 2. In principle, MAO can interact with the metal complex in three different ways (Scheme 2): by extracting the allyl group as in route a; by extracting the phosphine ligand as in route b; by coordinating through the oxygen atom to the cobalt atom, thus destabilizing in some way the cobalt complex, and favoring the insertion of the diene monomer at C3 of the allyl group, as shown in route c. According to route a, no metal-carbon bond for the monomer insertion is available; following route b no more ligand remains coordinated to the cobalt atom, meaning that we should not observe any influence of the ligand on the polymerization stereoselectivity, while, as reported above, 1,2 poly(1,3-butadiene)s having different tacticity, depending on the type of ligand coordinated to the cobalt atom, are obtained. It follows indeed that route c appears to be the most probable one. This hypothesis, in our opinion, is also in agreement with the fact that allene can be polymerized also in absence of MAO; allene could coordinate with only one double bond, thus favoring the insertion of the coordinated 1,3-butadiene, and then inserting itself into the allyl group of the growing chain. On the other hand, a similar behavior was observed with the allyl cobalt complex (η^4-C_4H_6)(η^5-C_8H_{13})Co, which was found to be extremely active in the polymerization of 1,3-butadiene to 1,2 polymers only if used in combination with CS_2 [12,15].

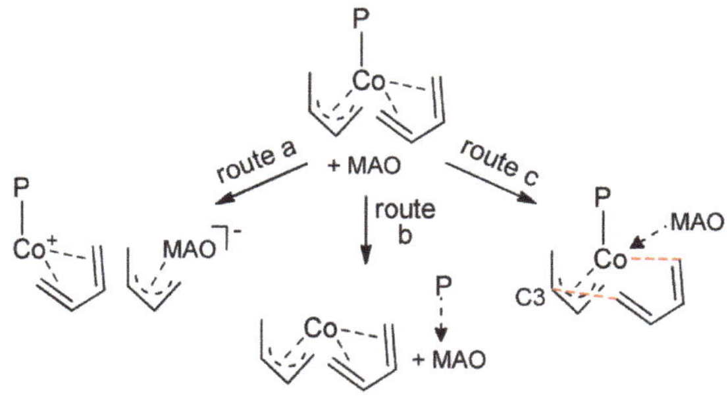

Scheme 2. Simplified illustration of the possible interactions between MAO and the allyl cobalt complexes.

The different behavior, regarding activity and stereospecificity, exhibited by the various cobalt complexes in the polymerization of the different diene monomers is even more difficult to interpret, but it is, in some way, in agreement with that observed in the polymerization of 1,3-dienes with the catalytic systems $CoCl_2(PRPh_2)_2$/MAO (R = methyl, ethyl *n*-propyl, *i*-propyl, cyclohexyl) [20,21]. We reported in previous papers that, in general, the stereoselectivity in the polymerization of 1,3-dienes with transition metal complex-based catalysts [e.g., $CrCl_2(L)$/MAO (L = bidentate phosphine); $FeCl_2(L)_2$-MAO (L = phenanthroline); $CoCl_2(L)_2$/MAO (L = monodentate phosphine)] was strongly affected by the catalyst structure (i.e., type of ligand bonded to the metal atom) and by the monomer structure [22–30]. The obtained results clearly indicated that (i) the *exo-exo* orientation of the allyl group of the growing chain and of the coordinated monomer was strongly favored in case of 1,3-butadiene and terminally substituted monomers, and of hindered ligands on the cobalt atom; (ii) the *exo-endo* orientation was instead favored in case of internally substituted 1,3-butadienes (e.g., isoprene and 3-methyl-1,3-pentadiene) and low hindered ligands on the metal atom [18]. Such a behavior is now confirmed, and even more evident, in case of the polymerization of 1,3-dienes with the allyl cobalt complexes object of the present paper. Complex **1**, exhibiting an *exo-exo* orientation of allyl group/coordinated monomer, is very active and selective only with terminally substituted monomers, that is monomers preferring such an orientation, while it is almost inactive with internally substituted monomers (isoprene and 3-methyl-1,3-pentadiene), that is monomers having higher affinity for the *exo-endo* orientation. The cobalt complexes **2** and **3**, for which the NMR analysis indicated the presence of both the orientations, *exo-endo* (probably the preferred one) and *exo-exo*, resulted instead scarcely active and selective with terminally substituted 1,3-butadienes, that is monomers preferring *exo-exo* orientations (e.g., 1,3-butadiene, 1,3-pentadiene and 1,3-hexadiene), while they were extremely active and selective in the polymerization of isoprene and 3-methyl-1,3-pentadiene, monomers preferring *exo-endo* orientations. Obviously, what is reported above has to be considered only a working hypothesis to provide a plausible interpretation for the obtained results and to stimulate discussion on this topic; further experimentation will be necessary and will be carried out to confirm or deny such an interpretation.

4. Materials and Methods

4.1. Materials

Anhydrous cobalt dichloride (Aldrich, St. Louis, MO USA, 99.9% pure), Zn powder (Strem, Newburyport, MA, USA, 99.9%), and methylaluminoxane (MAO) (Aldrich, 10 wt % solution in toluene) were used as received. Ethyl alcohol (Aldrich, 96%) was degassed under vacuum, then by bubbling dry dinitrogen and kept over molecular sieves; heptane (Aldrich, >99%) was refluxed over Na/K alloy for 8 h, then distilled and stored over molecular sieves under dry dinitrogen; toluene

(Aldrich, 99.5%) was refluxed over Na for 8 h, then distilled and stored over molecular sieves under dry nitrogen. 1,2 propadiene (Aldrich, ≥96%) was used as received. 1,3-Butadiene (Air Liquide, Paris, France, >99.5%) was evaporated from the container prior to each run, dried by passing through a column packed with molecular sieves and condensed into the reactor which had been precooled to −20 °C. Isoprene (Aldrich ≥99.5%), (*E*)-1,3-pentadiene (Aldrich, ≥96%), 1,3-hexadiene (Aldrich, 95%, mixture of (Z) and (E) isomers), and 3-methyl-1,3-pentadiene (Aldrich, 98%, mixture of (Z) and (E) isomers) were refluxed over calcium hydride for 3 h, then distilled trap-to-trap and stored under dry nitrogen. $CoCl_2(PRPh_2)_2$ (R = Me, Cy) were prepared as reported in literature [20,21,31].

4.2. Synthesis of Cobalt Complexes

4.2.1. Synthesis of $(\eta^3\text{-}C_4H_7)(\eta^4\text{-}C_4H_6)Co(PCyPh_2)$ (**1**)

The cobalt complex $CoCl_2(PCyPh_2)_2$ (2.53 g, 3.82 mmol) was introduced in a 100 mL round-bottomed Schlenk flask. Ethanol (3 mL), and a solution of 1,3-butadiene (12.4 mL, 159 mmol) in heptane (13 mL) were then added. The solution so obtained was cooled at −30 °C, and zinc powder (3 g, 45.9 mmol) was finally added. The suspension was kept under stirring at −30 °C for one day, then at −10 °C for one day, at 0 °C for a further day, and finally at +20 °C for a couple of days. The color of the final suspension was green, but the supernatant solution appeared red after decanting. The suspension was filtered, the red filtrate slightly concentrated under vacuum, and then cooled to −30 °C. After three days at low temperature, red single crystals, suitable for X-ray analysis, were formed, which were separated and dried under vacuum at room temperature. Yield: 0.80 g (47.9% based on $CoCl_2$). NMR and X-ray characterization of the complex is reported in the text.

4.2.2. Synthesis of $(\eta^3\text{-}C_4H_7)(\eta^4\text{-}C_4H_6)Co(PMePh_2)$ (**2**)

The cobalt complex $CoCl_2(PMePh_2)_2$ (2.89 g, 5.4 mmol) was introduced in a 100 mL round-bottomed Schlenk flask. Ethanol (3 mL), and a solution of butadiene (17 mL, 220 mmol) in heptane (17 mL) was then added. The solution so obtained was cooled at −30 °C, and the zinc powder (4.15 g, 63.5 mmol) was finally added. The suspension was kept at −30 °C under stirring for one day, then at 0 °C for one day and finally at +20 °C for a further day. The color of the suspension changed from blue to green to orange–yellow to the final red. After three days, the suspension was filtered; the filtrate was transferred in a refrigerator at −25 °C and kept at this temperature for three days. Red crystals of **2**, unfortunately not suitable for X-ray molecular structure determination, were formed, which were separated from the solution and dried under vacuum. Yield: 1.06 g (53.4% based on $CoCl_2$). NMR of the complex is reported in the text.

4.2.3. Synthesis of $(\eta^3\text{-}C_5H_9)(\eta^4\text{-}C_5H_8)Co(PMePh_2)$ (**3**)

The cobalt complex $CoCl_2(PMePh_2)_2$ (2.85 g, 5.4 mmol) was introduced in a 100 mL round-bottomed Schlenk flask. Ethanol (3 mL), heptane (17 mL) and isoprene (21.5 mL, 215 mmol) were then added. The solution so obtained was cooled at −30 °C, and zinc powder (4.17 g, 63.7 mmol) was finally added. The suspension was kept under stirring at −30 °C for one day, then at −10 °C for one day, and finally at +20 °C for one day. The color of the final suspension was red. The suspension was filtered, the red filtrate slightly concentrated under vacuum, and then cooled to −30 °C. No crystals were formed after three days; the solvent was indeed removed under vacuum, and pentane was added on the residue. The solution was filtered, and then kept at low temperature for crystallization. After one week, no precipitate was formed; the solvent was again removed under vacuum obtaining a highly viscous red oil as a residue which was dried under vacuum. Yield: 3.59 g (81.6% based on $CoCl_2$). NMR of the complex is reported in the text.

4.3. Cobalt Complexes Characterization

4.3.1. X-ray Crystallographic Studies of (η^3-C_4H_7)(η^4-C_4H_6)Co($PCyPh_2$) (**1**)

A summary of the experimental details concerning the X-ray diffraction study of complex **1** is reported in Tables 1 and S1. The intensity data were collected on a Bruker Smart Apex CCD area detector (Bruker AXS Inc., Madison, WI, USA) using graphite-monochromated Mo Kα radiation (λ = 0.71073 Å). Data reduction was made with SAINT and absorption corrections based on multiscan were obtained by SADABS [32]. The structure was solved by SHELXS-97 [33] and refined on F2 by full-matrix least-squares using SHELXL-14 [34]. All the non-hydrogen atoms were refined anisotropically. Hydrogen atoms were placed in calculated positions and included in the refinement as 'riding', excepted those bonded to the η^3 (C1–C3) and η^4 (C5–C8) coordinated carbon atoms. The H isotropic thermal parameters were fixed at 1.2 (1.5 for methyl groups) times the equivalent thermal parameter of the atoms to which they are bonded. The program ORTEPIII was used for graphics [35]. Crystallographic data were deposited with the Cambridge Crystallographic Data Centre as supplementary publication no. CCDC 1585813. These data can be obtained free of charge via www.ccdc.cam.ac.uk/conts/retrieving.html (or from CCDC, 12 Union Road, Cambridge CB2 1EZ, UK; fax: +44-1223-336-033; e-mail: deposit@ccdc.cam.ac.uk).

4.3.2. Computational Details

Geometry optimization of isolated, gas phase complexes (η^3-C_4H_7)(η^4-C_4H_6)Co($PRPh_2$) with R = cyclohexyl, phenyl and methyl were performed by the Gaussian16 program package [36], using the 6–311 g(d) basis set and the M06 functional, due to its specific parametrization on organometallic complexes [37]. The X-ray molecular structure of the complex with R = cyclohexyl was used as starting point for geometry optimization of all compounds, after the required atoms substitutions, owing to the lack of either the structure for the complex with R = methyl or the Cartesian coordinates of the deposited structure for the complex with R = phenyl [16].

4.3.3. NMR Characterization

^{1}H and 2D experiments were acquired on Bruker DMX 500 spectrometer, and on Bruker Avance 400 (Bruker Italia S.r.l., Milano, Italy). Samples were dissolved in hexadeuterobenzene, at 25 °C and were internally referred to the ^{1}H signal of residual benzene. The ^{1}H NMR conditions were the following: 5 mm probe; 8.75 μs as 90° pulse, relaxation delay 3 s; acquisition time 2.72 s. TOCSY (TOtal Correlation SpectroscopY) experiments correlate protons that interact through J-couplings showing correlations between all protons within a given spin system. The experiments were performed using a standard pulse sequence, with a mixing time of 0.1 s; power level for tocsy spin-lock 12 dB; low power pulse 52.80 μs. ROESY (Rotating frame Overhauser Effect SpectroscopY) experiments allow to detect protons that are close to each other in the space, even if they are not bonded. Spectra were recorded in phase sensitive mode using a standard pulse sequence; power level for roesy spin-lock 20 dB; spin-lock pulse 200 ms.

4.4. Polymerization

All operations were carried out under an atmosphere of dry dinitrogen. A standard procedure is reported. 1,3-Butadiene was condensed into a 25 mL dried glass reactor kept at −20 °C, then toluene was added and the solution so obtained was brought to the desired polymerization temperature. The cobalt allyl compound was then added as toluene solution. The polymerization was terminated with methanol containing a small amount of hydrochloric acid; the polymer was coagulated and repeatedly washed with methanol, then dried in vacuum at room temperature to constant weight. MAO, when used, was added just before the introduction of the cobalt compound.

The polymerizations with the other monomers (i.e., 1,2 propadiene, isoprene, (*E*)-1,3-pentadiene, 1,3-hexadiene, 3-methyl-1,3-pentadiene) were carried out in the same way.

4.5. Polymer Characterization

^{13}C and ^{1}H NMR measurements were carried out on a Bruker AVANCE 400 spectrometer, operating at 400 MHz for ^{1}H and 100.58 MHz for ^{13}C. The spectra were obtained in $C_2D_2Cl_4$ at 103 °C (hexamethyldisiloxane, HMDS, as internal standard). The concentration of polymer solutions was about 10 wt %. ^{13}C NMR conditions were the following: 10 mm probe; 14.50 μs as 90° pulse, relaxation delay 18 s; acquisition time 1.87 s. Proton broad-band decoupling was achieved with a 1D sequence using bi_waltz_16_32 power-gated decoupling. DSC scans were carried out on a Perkin Elmer Pyris 1 Instrument (Perkin Elmer, Milano, Italy) equipped with a liquid nitrogen sub-ambient device. The sample, ca. 4 mg, was placed in a sealed aluminum pan and the measurements were carried out using heating and cooling rates of 20 °C/min. The molecular weight averages (M_w) and the molecular weight distribution (M_w/M_n) were obtained by a Waters GPCV 2000 system (Waters S.p.A., Milano, Italy) using two on-line detectors: a differential viscosimeter and a refractometer. The experimental conditions consisted of two PLgel Mixed C columns, *ortho*-dichlorobenzene as mobile phase, 0.8 mL/min of flow rate and 145 °C of temperature. The calibration of the GPC system was carried out using eighteen narrow M_w/M_n polystyrene standards with the molar weights ranging from 162 to 5.6×10^6 g/mol. X-ray powder diffraction profiles were obtained with Ni filtered CuKα radiation with an automatic diffractometer X-Pert by Panalytical (Panalytical, Almelo, The Netherlands).

5. Conclusions

Novel allyl cobalt phosphine complexes were synthesized, which were found to be able to polymerize 1,3-dienes and allene, exhibiting different activity and selectivity depending on the type of phosphine ligand coordinated to the metal atom and the type of monomer. The structures of these new allyl complexes, determined through NMR and X-ray techniques, and the structure of the polymers obtained, determined by NMR spectroscopy, provided further support and confirmation of the 1,3-dienes polymerization mechanism that has been proposed previously [18,22].

Supplementary Materials: The following are available online at www.mdpi.com/2073-4344/7/12/381/s1, Table S1: Crystallographic data and results of structure refinement for **1**. Figure S1: TOCSY NMR spectrum of **2**. Figure S2: TOCSY NMR spectrum of **3**. Figure S3: ROESY NMR spectrum of **3**. Figure S4: ROESY NMR spectrum of **2**. This material is available free of charge via the Internet at http://sciencedirect.com.

Author Contributions: Giovanni Ricci conceived and designed the experiments, performed the organometallic syntheses, interpreted the results and wrote the paper. Giuseppe Leone performed the organometallic syntheses and the polymerizations. Antonella Caterina Boccia performed the NMR characterization of the cobalt complexes and contributed to write the paper. Alessandra Forni determined the X-ray molecular structures of the complexes, performed the quantum-mechanical calculations and contributed to write the paper.

Conflicts of Interest: The authors declare no conflict of interest.

References

1. Porri, L.; Giarrusso, A. Part II: Conjugated Diene Polymerization. In *Comprehensive Polymer Science*; Pergamon: Oxford, UK, 1989; Volume 4, pp. 53–108.
2. Thiele, S.K.H.; Wilson, D.R. Alternate Transition Metal Complex Based Diene Polymerization. *J. Macromol. Sci.* **2003**, *C43*, 581–628. [CrossRef]
3. Osakada, K.; Takeuchi, D. Coordination Polymerization of Dienes, Allenes, and Methylenecycloalkanes. *Adv. Polym. Sci.* **2004**, *171*, 137–194.
4. Friebe, L.; Nuyken, O.; Obrecht, W. Neodymium-Based Ziegler/Natta Catalysts and their Application in Diene Polymerization. *Adv. Polym. Sci.* **2006**, *204*, 1–154.
5. Dagorne, S.; Fliedel, C. Organoaluminum species in homogeneous polymerization catalysis. *Top. Organomet. Chem.* **2013**, *41*, 125–172.

6. Wilke, G.; Bogdanovic, B.; Hardt, P.; Heimbach, P.; Keim, W.; Kromer, M.; Oberkirch, W.; Tanaka, K.; Steinrucke, E.; Walter, D.; et al. Allyl-Transition Metal Systems. *Angew. Chem. Int. Ed. Engl.* **1966**, *5*, 151–164. [CrossRef]
7. Porri, L.; Natta, G.; Gallazzi, M.C. Polymerization of butadiene and cycloolefins by π-allylnickel bromide. *Chim. Ind. (Milan)* **1964**, *46*, 428–429, (*Chem. Abstr.* **1964**, *61*, 1944).
8. Shmonina, V.L.; Stefanovskaya, N.N.; Tinyakova, E.I.; Dolgoplosk, B.A. Stereospecific polymerization of dienes catalyzed by π-allyl complexes of chromium on aluminosilicate. *Vysokomol. Soedin. Ser. B* **1970**, *12*, 566.
9. Teyssie, P.; Dawans, F. π-Allyl-type polymerization. *Ind. Eng. Chem. Prod. Res. Dev.* **1971**, *10*, 261–269.
10. Dolgoplosk, B.A.; Tinyakova, E.I. Nature of active centers and mechanism of coordination polymerization. I. Nature of active centers during the stereospecific polymerization of dienes and mechanism of stereoregulation. *Vysokomol. Soedin. Ser. A* **1977**, *19*, 2441–2463.
11. Ashitaka, H.; Ishikawa, H.; Ueno, H.; Nagasaka, A. Syndiotactic 1,2-polybutadiene with Co-CS_2 catalyst system. I. Preparation, properties, and application of highly crystalline syndiotactic 1,2-polybutadiene. *J. Polym. Sci. Polym. Chem. Ed.* **1983**, *21*, 1853–1860. [CrossRef]
12. Ashitaka, H.; Jinda, K.; Ueno, H. Syndiotactic 1,2-polybutadiene with Co-CS_2 catalyst system. II. Catalysts for stereospecific polymerization of butadiene to syndiotactic 1,2-polybutadiene. *J. Polym. Sci. Polym. Chem. Ed.* **1983**, *21*, 1951–1972. [CrossRef]
13. Ashitaka, H.; Inaishi, K.; Ueno, H. Syndiotactic 1,2-polybutadiene with Co-CS_2 catalyst system. III. ^{1}H- and ^{13}C-NMR study of highly syndiotactic 1,2-polybutadiene. *J. Polym. Sci. Polym. Chem. Ed.* **1983**, *21*, 1973–1988. [CrossRef]
14. Ashitaka, H.; Jinda, K.; Ueno, H. Syndiotactic 1,2-polybutadiene with Co-CS_2 catalyst system. IV. Mechanism of syndiotactic polymerization of butadiene with cobalt compounds–organoaluminum–CS_2. *J. Polym. Sci. Polym. Chem. Ed.* **1983**, *21*, 1989–1995. [CrossRef]
15. Ricci, G.; Italia, S.; Porri, L. Polymerization of butadiene to 1,2-syndiotactic polymer with $(\eta^5\text{-}C_8H_{13})(\eta^4\text{-}C_4H_6)Co$. Some observations on the factors that determine the stereospecificity. *Polym. Commun.* **1988**, *29*, 305–307.
16. Porri, L.; Vitulli, G.; Zocchi, M.; Allegra, G. A but-2-enyl-buta-1,3-diene complex of cobalt. *J. Chem. Soc. D* **1969**, *6*, 276–277. [CrossRef]
17. Vitulli, G.; Porri, L.; Segre, A.L. Preparation and Properties of a But-2-enyl Butadiene Complex of Cobalt and of an Analogous Complex from Isoprene. *J. Chem. Soc. A* **1971**, 3246–3250. [CrossRef]
18. Ricci, G.; Sommazzi, A.; Masi, F.; Ricci, M.; Boglia, A.; Leone, G. Well Defined Transition Metal Complexes with Phosphorus and Nitrogen Ligands for 1,3-Dienes Polymerization. *Coord. Chem. Rev.* **2010**, *254*, 661–676. [CrossRef]
19. Ricci, G.; Leone, G.; Boglia, A.; Boccia, A.C.; Zetta, L. *cis*-1,4-*alt*-3,4 Polyisoprene: Synthesis and Characterization. *Macromolecules* **2009**, *42*, 9263–9267. [CrossRef]
20. Ricci, G.; Forni, A.; Boglia, A.; Motta, T.; Zannoni, G.; Canetti, M.; Bertini, F. Synthesis and X-ray Structure of $CoCl_2(P^iPrPh_2)_2$. A New Highly Active and Stereospecific Catalyst for 1,2 Polymerization of Conjugated Dienes When Used in Association with MAO. *Macromolecules* **2005**, *38*, 1064–1070. [CrossRef]
21. Ricci, G.; Forni, A.; Boglia, A.; Sommazzi, A.; Masi, F. Synthesis, structure and butadiene polymerization behavior of $CoCl_2(PR_xPh_{3-x})_2$ (R = methyl, ethyl, propyl, allyl, isopropyl, cyclohexyl; $x = 1, 2$). Influence of the phosphorous ligand on polymerization stereoselectivity. *J. Org. Chem.* **2005**, *690*, 1845–1854. [CrossRef]
22. Porri, L.; Giarrusso, A.; Ricci, G. Recent views on the mechanism of diolefin polymerization with transition metal initiator systems. *Prog. Polym. Sci.* **1991**, *16*, 405–441. [CrossRef]
23. Ricci, G.; Battistella, M.; Porri, L. Chemoselectivity and Stereospecificity of Chromium(II) Catalysts for 1,3-Diene Polymerization. *Macromolecules* **2001**, *34*, 5766–5769. [CrossRef]
24. Ricci, G.; Morganti, D.; Sommazzi, A.; Santi, R.; Masi, F. Polymerization of 1,3-dienes with iron complexes based catalysts. Influence of the ligand on catalyst activity and stereospecificity. *J. Mol. Catal. A Chem.* **2003**, *204/205*, 287–293. [CrossRef]
25. Ricci, G.; Forni, A.; Boglia, A.; Sonzogni, M. New Chromium(II) Bidentate Phosphine Complexes: Synthesis, Characterization, and Behavior in the Polymerization of 1,3-Butadiene. *Organometallics* **2004**, *23*, 3727–3732. [CrossRef]
26. Ricci, G.; Forni, A.; Boglia, A.; Motta, T. Synthesis, structure, and butadiene polymerization behavior of alkylphosphine cobalt(II) complexes. *J. Mol. Catal. A Chem.* **2005**, *226*, 235–241. [CrossRef]

27. Ricci, G.; Boglia, A.; Motta, T. Synthesis of new Cr(II) complexes with bidentate phosphine ligands and their behavior in the polymerization of butadiene. Influence of the phosphine bite angle on catalyst activity and stereoselectivity. *J. Mol. Catal. A Chem.* **2007**, *267*, 102–107. [CrossRef]
28. Ricci, G.; Leone, G.; Boglia, A.; Bertini, F.; Boccia, A.C.; Zetta, L. Synthesis and Characterization of Isotactic 1,2-Poly(*E*-3-methyl-1,3-pentadiene). Some Remarks about the Influence of Monomer Structure on Polymerization Stereoselectivity. *Macromolecules* **2009**, *42*, 3048–3056. [CrossRef]
29. Boccia, A.C.; Leone, G.; Boglia, A.; Ricci, G. Novel stereoregular *cis*-1,4 and *trans*-1,2 poly(diene)s: Synthesis, characterization, and mechanistic considerations. *Polymer* **2013**, *54*, 3492–3503. [CrossRef]
30. Ricci, G.; Leone, G. Recent progresses in the polymerization of butadiene over the last decade. *Polyolefins J.* **2014**, *1*, 43–60.
31. Chatt, J.; Shaw, B.L. Alkyls and aryls of transition metals. Part IV. Cobalt(II) and iron(II) derivatives. *J. Chem. Soc.* **1961**, 285–290. [CrossRef]
32. Bruker. *SMART, SAINT and SADABS*; Bruker AXS Inc.: Madison, WI, USA, 1997.
33. Sheldrick, G.M. A short history of SHELX. *Acta Cryst.* **2008**, *A64*, 112–122. [CrossRef] [PubMed]
34. Sheldrick, G.M. Crystal structure refinement with SHELXL. *Acta Cryst.* **2015**, *C71*, 3–8.
35. Burnett, M.N.; Johnson, C.K. *ORTEP-III: Oak Ridge Thermal Ellipsoid Plot Program for Crystal Structure Illustrations*; Oak Ridge National Laboratory Report ORNL-6895; Oak Ridge National Laboratory: Oak Ridge, TN, USA, 1996.
36. Frisch, M.J.; Trucks, G.W.; Schlegel, H.B.; Scuseria, G.E.; Robb, M.A.; Cheeseman, J.R.; Montgomery, J.A., Jr.; Vreven, T.; Kudin, K.N.; Burant, J.C. *Gaussian 16, Revision A.03*; Gaussian, Inc.: Wallingford, CT, USA, 2016.
37. Zhao, Y.; Truhlar, D.G. The Mo6 suite of density functionals for main group thermochemistry, thermochemical kinetics, noncovalent interactions, excited states, and transition elements: Two new functionals and systematic testing of four Mo6-class functionals and 12 other functionals. *Theor. Chem. Acc.* **2008**, *120*, 215–241.

catalysts

MDPI

Article

Synthesis, Structure and 1,3-Butadiene Polymerization Behavior of Vanadium(III) Phosphine Complexes

Giuseppe Leone [1], Giorgia Zanchin [1,2], Ivana Pierro [1,3], Anna Sommazzi [4], Alessandra Forni [5] and Giovanni Ricci [1,*]

1 CNR, Istituto per lo Studio delle Macromolecole (ISMAC), via A. Corti 12, I-20133 Milano, Italy; giuseppe.leone@ismac.cnr.it (G.L.); giorgia.zanchin@ismac.cnr.it (G.Z.); ivana.pierro@ismac.cnr.it (I.P.)
2 Dipartimento di Chimica, Università degli Studi di Milano, via C. Golgi 19, I-20133 Milano, Italy
3 Dipartimento di Scienze Chimiche, Università degli Studi di Napoli Federico II, Complesso Monte S. Angelo, Via Cintia, I-80126 Napoli, Italy
4 Versalis S.p.A., Istituto Eni Donegani, via G. Fauser 4, I-28100 Novara, Italy; anna.sommazzi@versalis.eni.com
5 CNR, Istituto di Scienze e Tecnologie Molecolari (ISTM), Università degli Studi di Milano, via C. Golgi 19, I-20133 Milano, Italy; alessandra.forni@istm.cnr.it
* Correspondence: giovanni.ricci@ismac.cnr.it; Tel.: +39-02-2369-9376

Received: 9 November 2017; Accepted: 25 November 2017; Published: 28 November 2017

Abstract: A series of vanadium(III) complexes bearing monodentate tertiary phosphine ligands of the type $VCl_3(PR_nPh_{3-n})_2$ (n = 0 (**1a**); n = 1 and R = Me (**1b**), Et (**1c**), ^{i}Pr (**1d**), Cy (**1e**); n = 2 and R = Me (**1f**), Et (**1g**), Cy (**1h**)), and $VCl_3(PR_3)_2$ (R = Cyp (**2a**), Cy (**2b**), ^{n}Pr (**2c**), ^{t}Bu (**2d**)) were synthesized and characterized. In the case of **1c**, **1g** and **2a** single crystals were also obtained and their molecular structures were determined. All the complexes were used, in combination with methylaluminoxane (sMAO) or $AlMe_3$–free MAO (dMAO), for the polymerization of 1,3-butadiene, exhibiting rather good activity and giving polymers with different microstructure depending on the nature of the phosphine ligand and the type of co-catalyst employed.

Keywords: vanadium phosphine complexes; X-ray structure; 1,3-butadiene polymerization; poly(1,3-butadiene)

1. Introduction

The polymerization of 1,3-butadiene plays an important role in the field of stereospecific polymerization, and it is of great interest for both the academic and industrial world. Lanthanides (e.g., Nd, Pr) and transition metals (e.g., Ti, V, Cr, Fe, Co, Ni) compounds in combination with suitable alkylating agents (e.g., $AlEt_3$, $AlEt_2Cl$, MAO) were successfully used for the polymerization of 1,3-butadiene, affording polymers with different structure (i.e., *trans*-1,4, *cis*-1,4, isotactic 1,2, syndiotactic 1,2 and mixed *cis*-1,4/1,2 structure with variable 1,2 content) depending on the catalyst nature [1–3].

In particular, vanadium based catalysts are mainly known for their ability to give highly *trans*-1,4 poly(1,3-butadiene) [1]. A crystalline, highly stereoregular *trans*-1,4 poly(1,3-butadiene) (*trans*-1,4 $\geq$ 99%) with high molecular weight and melting point of about 130 °C was synthesized with the heterogeneous systems obtained by combining a vanadium chloride (e.g., VCl_3, VCl_4) with an aluminum alkyl (e.g., $AlEt_3$, $AlEt_2Cl$) [4]. A highly *trans*-1,4 poly(1,3-butadiene), but with lower molecular weight and melting point, was also obtained with the soluble catalytic systems $VCl_3{\cdot}3THF/AlEt_2Cl$ [5], $V(acac)_3/AlEt_2Cl$ (acac = acetylacetonate) [6], and $V(acac)_3$/MAO [7,8]. *trans*-1,4 Poly(1,3-butadiene)s were also obtained with catalysts based on vanadium(III) complexes bearing a bis(imino)pyridyl ligand (i.e., $VCl_3\{2,6\text{-bis}[(2,6\text{-}i\text{Pr}_2C_6H_3)NC(Me)]_2(C_5H_3N)\}$); the activity of these systems was, however, rather low [9]. Some vanadium based catalysts were also able to give 1,2 syndiotactic and predominantly *cis*-1,4 poly(1,3-butadiene)s. V(acac)$_3$, in combination with AlEt$_3$, was found to be an efficient catalyst for the synthesis of syndiotactic 1,2 poly(butadiene)s [10]; V(η^5-C$_5$H$_5$)$_2$Cl and V(η^5-C$_5$H$_4$Me)(PEt$_3$)$_2$Cl$_2$, in combination with MAO, gave polymers with a predominantly *cis*-1,4 structure (*cis* content of about 85%) with a quite good activity [11–13]. Essentially *cis*-1,4 poly(1,3-butadiene)s were also obtained by using some amino-functionalized cyclopentadienyl vanadium(III) complexes (e.g., V{η^5-C$_5$H$_4$CH(CH$_2$)$_4$NMe}$_2$Cl, V{η^5-C$_5$H$_4$(CH$_2$)$_2$N(CH$_2$)$_5$}$_2$Cl) [14].

In recent years, several studies on the use of various transition metal phosphine complexes for the polymerization of 1,3-dienes have been reported [15–24]. Catalysts obtained by combining various phosphine complexes of chromium(II) [15–17] and cobalt(II) [18–20] with MAO were found to be extremely active and selective in the polymerization of 1,3-dienes: highly stereoregular polymers having different structure were obtained depending on the type of metal, the type of ligand and the monomer polymerized [21–24]. As far as we know, up to now, only few examples of vanadium(III) phosphine complexes were reported in the literature [25–31], and only some of them have been used for the polymerization of 1,3-butadiene [31].

In this paper, we report on the synthesis and characterization of several vanadium(III) complexes of the type VCl$_3$(PR$_n$Ph$_{3\text{-}n}$)$_2$ (n = 0 (**1a**); n = 1 and R = Me (**1b**), Et (**1c**), ^{i}Pr (**1d**), Cy (**1e**); n = 2 and R = Me (**1f**), Et (**1g**), Cy (**1h**), and VCl$_3$(PR$_3$)$_2$ (R = cyclopentyl (Cyp, **2a**), Cy (**2b**), ^{n}Pr (**2c**), ^{t}Bu (**2d**), including single-crystal X-ray structural determination of **1c**, **1g** and **2a**. A preliminary study investigating the catalytic behavior of these novel complexes in the polymerization of 1,3-butadiene is presented.

2. Results and Discussion

The study of phosphine complexes of transition metal halides is an important topic and includes different compounds depending on the metal oxidation state and its position in the periodic table. Toward the left side (i.e., Group 4 and 5) and with oxidation states of III and IV, the phosphine–metal bonds tend to be relatively long, weak and labile, and they have little if any π character [32], even if there are examples for which bond lengths do not correlate with their expected bond orders, as in some titanocenes phosphines adducts [33].

Among the metal–phosphine complexes reported, our interest was in preparing vanadium(III) complexes. Complexes of vanadium(III) with tertiary phosphines were prepared from VCl$_3$ or VCl$_3$·3THF. It has been reported that PPh$_3$ and PCy$_3$ reacted with VCl$_3$ to give dimers, i.e., [VCl$_3$(PR$_3$)]$_2$ [29]. Mononuclear complexes with a coordination number of 5 were instead obtained by reacting VCl$_3$·3THF with PMePh$_2$, PEt$_2$Ph, and PR$_3$ (R = Me, Et); with a coordination number of 6 [i.e., VCl$_3$(P^{n}Pr)$_3$ and VCl$_3$(PR$_2$Ph)$_3$ (R = Me, ^{n}Pr)] by reacting VCl$_3$·3THF with P^{n}Pr$_3$, PMe$_2$Ph and P^{n}Pr$_2$Ph [26,28]. With other aliphatic phosphines, only intractable oils were obtained likely due to a reduction-hydrolysis process occurring during the reaction and subsequent workup [32].

In this work, we explore the chemical reactivity of VCl$_3$·3THF versus tertiary phosphine by extending the spectrum of phosphine ligands; The synthesis of vanadium complexes with PR$_n$Ph$_{3\text{-}n}$ (n = 0, 1, 2; R = Me, Et, ^{i}Pr, Cy) and PR$_3$ (R = Cyp, Cy, ^{n}Pr, ^{t}Bu) is reported, including single-crystal X-ray structural determination for VCl$_3$(PEtPh$_2$)$_2$ (**1c**), VCl$_3$(PEt$_2$Ph)$_2$ (**1g**) and VCl$_3$(PCyp$_3$)$_2$ (**2a**).

2.1. Synthesis and Structure of V(III)–Phosphine Complexes

Complexes of type VCl$_3$(PR$_n$Ph$_{3\text{-}n}$)$_2$ (n = 0; n = 1, R = Me, Et, Cy; n = 2, R = Me) and VCl$_3$(PR$_3$)$_2$ (R = Cyp, Cy) were obtained in good yield as microcrystalline solids by addition of the phosphine to a VCl$_3$·3THF suspension in toluene (P/V = 4:1 mole ratio). Single crystals, suitable for X-ray structure determination, were obtained for **1b**, whose structure was previously reported [25,27,30], **1c** and **2a** from cold pentane. In contrast, the reaction in toluene with phosphine of type PR$_n$Ph$_{3\text{-}n}$ (n = 2, R = Et, ^{t}Bu, Cy) and PR$_3$ (R = ^{n}Pr, ^{t}Bu) gave an oily product, difficult to recover. Dusty solids were instead successfully obtained by performing the reaction in THF and adding the ligand dropwise; single crystals, suitable for X-ray structure determination, were obtained only for **1g**. Due to their low stability in air, all crystals required low temperature to minimize crystal decay during X-ray data collection (130 K for **1b** [30], and **1c**, 150 K for **1g** and 100 K for **2a**, according to their different stability). A list of selected bond distances and angles for **1b** [30], **1c**, **1g** and **2a** as obtained by X-ray investigation is reported in Table 1, while a view of the molecular structures of **1c**, **1g** and **2a** is reported in Figure 1. DFT gas-phase geometry optimizations, starting from the experimental structures, have been performed on the four complexes in order to discern the effect of crystal packing, if any, on the experimental geometries. Two exchange-correlation functionals have been used to assess the reliability of the theoretical investigation: M06 [34], due to its specific parametrization on organometallic complexes, and B3LYP [35–37], by far the most used functional for geometry optimizations. The results are reported as well in Table 1. At a glance, M06 showed much better performance with respect to B3LYP as far as bond lengths are concerned, while bond angles are in all cases quite similar.

Table 1. Selected X-ray (first row) and Computed (UM06/6-311g(d) and UB3LYP/6-311g(d), second and third row, respectively, in italics) Bond Lengths (Å) and Angles (°) for **1b** [30], **1c**, **1g** and **2a** [a].

Bond Lengths and Angle	$VCl_3(PMePh_2)_2$ (1b)	$VCl_3(PEtPh_2)_2$ (1c)	$VCl_3(PEt_2Ph)_2$ (1g)	$VCl_3(PCyp_3)_2$ (2a)
V–Cl	2.2287(8)	2.2408(6)	2.246(10)	2.247(10)
	2.243	*2.243*	*2.245*	*2.251*
	2.255	*2.255*	*2.258*	*2.261*
V–P	2.5280(6)	2.5465(6)	2.5196(11)	2.5696(10)
	2.537	*2.551*	*2.549*	*2.587*
	2.607	*2.617*	*2.613*	*2.660*
$P–C_{ar}$	1.820(2)	1.8251(19)	1.813(4)	-
	1.825	*1.826*	*1.828*	-
	1.840	*1.841*	*1.841*	-
$P–C_{aliph}$	1.822(2)	1.8332(19)	1.828(4)	1.847(3)
	1.828	*1.844*	*1.842*	*1.850*
	1.843	*1.861*	*1.860*	*1.869*
Cl–V–Cl	119.98(3)	119.99(2)	114.6(2)	120.00(4)
	120.00	*120.00*	*119.98*	*120.00*
	120.00	*120.00*	*119.98*	*120.00*
P–V–P	169.02(2)	177.87(2)	167.24(4)	170.48(3)
	174.74	*177.79*	*174.06*	*174.66*
	175.06	*177.78*	*174.25*	*174.77*
$C_{ar}–P–C_{ar}$	103.73(10)	103.85(8)	-	-
	103.64	*104.39*	-	-
	103.74	*104.29*	-	-
$C_{ar}–P–C_{aliph}$	105.20(11)	105.42(9)	105.1(2)	-
	104.03	*104.31*	*104.20*	-
	104.09	*104.53*	*104.17*	-
$C_{aliph}–P–C_{aliph}$	-	-	103.7(2)	105.5(2)
	-	-	*102.79*	*104.09*
	-	-	*102.87*	*104.99*

[a] Each reported value was obtained as weighted average on all the corresponding parameters present in the structure.

More precisely, both functionals provide similar trends in the geometrical parameters along the series of complexes, generally in agreement with the experimental ones. M06, however, reproduces bond lengths within 0.015 Å, with the only exception of V–P distances which show a maximum discrepancy of 0.029 Å. B3LYP overestimates bond lengths by up to 0.032 Å, excepted for the V–P bonds for which the maximum deviation grows up to 0.093 Å. In the following we therefore rely only on the M06 optimized geometry. The complexes adopt slightly distorted trigonal-bipyramidal geometry, where the main distortion comes from the deviation (by up to 13°) of the P–V–P angles from 180°. As previously discussed [30], such a deviation is mainly due to crystal packing forces (essentially C–H···Cl [38], C–H···π and π–π intermolecular interactions, the latter two arising in structures where phenylphosphines are present), as denoted by the more regular conformation as obtained by gas-phase calculations, which provide a maximum 6° deviation from a P–V–P straight angle. As a general observation, the V–P bond lengths, ranging on average from 2.520(1) to 2.570(1) Å (2.537 to 2.587 Å from UM06 calculations) in the compounds here investigated, appear to be much longer than the Mt–P (Mt = metal) ones as found in tetrahedral $MtCl_2(PR_nPh_{3-n})_2$ complexes, suggesting weaker bonds with scarce or null π-character for the vanadium complexes. For example, in the extended series of $CoCl_2(PR_nPh_{3-n})_2$ complexes previously reported by some of us [18–20,39], the longest Co–P bond distance was found to be 2.437(6) Å for the more sterically hindered $CoCl_2(PCy_3)_2$ complex [19], well below the V–P shortest distance here reported. Along the series of **1b**, **1c**, **1g** and **2a** complexes, the trend in the V–P and, in a lesser extent, the V–Cl average bond lengths generally follows the combined effect of the bulkiness of the phosphine ligand and its π-character. In fact, they increase with either increasing the hindrance of the phosphine and/or decreasing the π-character of the phosphine substituents. For example, going from **1b** to **1c**, the bulkier ethyl compared to the methyl group implies an increase in the average V–P distance, Δr(V–P), equal to 0.018(1) Å (0.014 Å from theory). Comparing **1b**, **1c** and **1g** with **2a**, we observe a definite increase Δr(V–P) = 0.042(1), 0.023(1) and 0.050(1) Å, respectively (0.050, 0.036 and 0.038 Å from theory), due to both the large hindrance of the cyclopentyl group and the absence of any π-electron group in the phosphine. Interestingly, replacement of one phenyl by an ethyl group (compare **1c** with **1g**), implying conflicting effects (reduced hindrance together with reduced π-character of the phosphine) leads to V–P shortening, Δr(V–P) = −0.027(1), suggesting a dominance of the steric over the electronic effect. However, this result should be also ascribed to crystal packing forces because gas-phase calculations provide only minor contraction, Δr(V–P) = −0.002 Å.

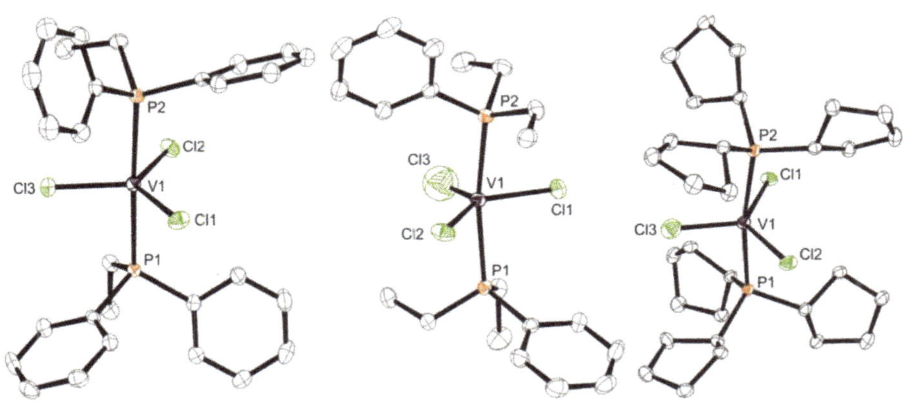

Figure 1. Molecular structures of $VCl_3(PEtPh_2)_2$ (**1c**) (**left**), $VCl_3(PEt_2Ph)_2$ (**1g**) (**center**) and $VCl_3(PCyp_3)_2$ (**2a**) (**right**) with thermal ellipsoids drawn at 50% probability level. Hydrogen atoms omitted for clarity.

Considering the P–C bond lengths involving the aromatic carbon atoms, P–C_{ar}, they are approximately constant within the experimental error, while those involving the aliphatic carbon atoms, P–C_{aliph}, follow the same trend as observed for the V–P bonds, though with a lower variability. Comparing **1b**, **1c** and **1g** with **2a**, we observe in fact an increase Δr(P–C_{aliph}) = 0.025(3), 0.014(3) and 0.019(4) Å, respectively (0.022, 0.006 and 0.008 Å from theory). The C–P–C bond angles are essentially unvaried in the four structures, with the only exception of the C_{aliph}–P–C_{aliph} angle, which as expected undergoes a small but non negligible increase, equal to 1.8(2)°, well reproduced by calculations (1.3°), going from **1g** to **2a**.

2.2. Polymerization of 1,3-Butadiene

The polymerization of 1,3-butadiene with all the vanadium(III) phosphine complexes synthesized, in the presence of white solid MAO [dMAO, prepared by removing toluene and $AlMe_3$ from commercially available MAO (sMAO)] or sMAO as co-catalyst, was examined. The polymerization runs were carried out in toluene at room temperature. The most significant results are summarized in Table 2.

All the vanadium complexes, in combination with dMAO, were active in the polymerization of 1,3-butadiene, giving low to good polymer yields, ranging from 13 to 57% (Table 2, entry 17 and 1, respectively), within the first 2 h of polymerization. By extending the polymerization runs up to 24 h, it was possible to reach complete monomer consumption (Table 2, entry 2), meaning that the catalysts are not deactivated with time, as it occurs in the polymerization of ethylene where high concentration of reoxidant (typically $Cl_3CCOOEt$ or chlorinated hydrocarbons) is required to keep the catalyst active [30,40,41]. This can be ascribed to the higher nucleophilicity of 1,3-butadiene with respect to a simple olefin, but mainly to the higher stability of the vanadium–π-allyl bond compared to the vanadium–alkyl bond of the intermediate species involved in the polymerization of 1,3-butadiene and ethylene, respectively [42].

The results in Table 2 are presented as a function of the phosphine donor ability [as measured by the electronic parameter υ_{CO} based on the carbonyl stretching frequencies in $Ni(CO)_3L$ complexes (L = phosphine ligand)] and steric properties (measured by cone angle, θ) as defined by Tolman [43,44], to highlight the influence of the ligand (i.e., the type of substituents on the phosphorous atoms) on the catalytic activity and selectivity. A general increasing of the activity by decreasing the σ-donor character (increasing π-acceptor character) of the phosphine ligand was observed, irrespective of the ligand steric hindrance. Electron-withdrawing groups on the ligand framework seem to enhance catalytic performance. Complexes with aryl phosphine exhibited higher activity than those bearing alkyl phosphines, as shown in Figure 2 (catalyst using $PCyp_3$ makes an exception to this trend). The P^tBu_3 (the strongest σ-donor) yields the least active system, confirming the strong influence of the phosphine basicity on the catalytic activity.

Table 2. Polymerization of 1,3-butadiene with the phosphine V(III) complexes [a].

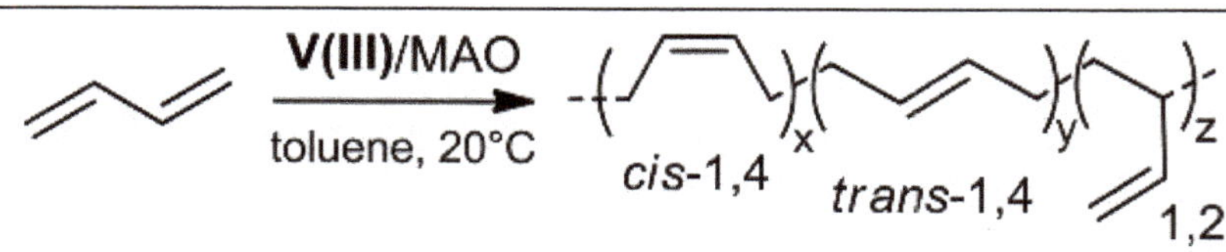

Entry	Catalyst	Phosphine (Type)	υ_{CO} [b] (cm^{-1})	θ [c] (°)	Yield (%)	N [d] (h^{-1})	*cis*:*trans*:1,2 [e] (mol %)	M_w [f] ($\times 10^3$)	M_w/M_n [f]
1	**1a**/dMAO	PPh_3	2068.9	145	57	739	33:31:36	275	1.9
2 *	**1a**/dMAO	PPh_3	2068.9	145	100	123	30:34:36	290	1.6
3 *	**1a**/sMAO	PPh_3	2068.9	145	51	63	65:19:16	177	4.2
4	**1b**/dMAO	$PMePh_2$	2067.0	136	46	593	29:37:34	270	2.1
5	**1b**/sMAO	$PMePh_2$	2067.0	136	7	93	69:17:14	164	1.9
6	**1c**/dMAO	$PEtPh_2$	2066.7	140	42	360	30:38:32	83	2.1
7	**1d**/dMAO	$P\textit{i}PrPh_2$	2065.7	150	50	648	28:33:39	112	3.5
8	**1d**/sMAO	$P\textit{i}PrPh_2$	2065.7	150	17	222	66:18:16	102	2.6
9	**1e**/dMAO	$PCyPh_2$	2064.8	153	46	593	37:28:35	116	3.0
10 *	**1e**/sMAO	$PCyPh_2$	2064.8	153	43	46	67:15:18	100	2.9

Table 2. *Cont.*

11	**1f**/dMAO	PMe_2Ph	2065.3	122	43	556	41:20:39	75	3.8
12	**1g**/dMAO	PEt_2Ph	2063.7	136	37	480	36:24:38	97	2.7
13	**1h**/dMAO	PCy_2Ph	2060.6	159	25	324	14:73:13	274	2.8
14	**2a**/dMAO	$PCyp_3$			36	472	46:33:21	183	3.5
15	**2b**/dMAO	PCy_3	2056.4	170	21	269	34:46:20	193	3.3
16	**2c**/dMAO	$P\mathit{n}Pr_3$	2059.1	132	26	333	36:21:43	296	1.8
17	**2d**/dMAO	$P\mathit{t}Bu_3$	2056.1	182	13	167	45:34:21	253	3.1

[a] polymerization conditions: 1,3-butadiene, 2 mL (1.4 g); solvent, toluene (16 mL); V(III)-complex, 10 μmol; Al/V, 1000; time, 2 h (* 24 h in runs 2, 3 and 10); [b] measure of the electron-donating properties of phosphine ligands, as reported by Tolman [43,44]; [c] phosphine cone angle, as reported by Tolman [43,44]; [d] N = moles of 1,3-butadiene polymerized per mol of vanadium per hour; [e] determined by ^1H and ^{13}C NMR; [f] determined by SEC.

The type of phosphine ligand was also found to have some effect, though not very pronounced, on the polymerization selectivity, as suggested by the microstructure of the resulting poly(1,3-butadiene)s determined by ^1H and ^{13}C NMR (Table 2 and Figures S1–S4). Generally, all the vanadium complexes, in combination with dMAO, gave polymers with a mixed *cis*-1,4/*trans*-1,4/1,2 structure, the 1,4 and 1,2 units being randomly distributed along the polymer chain, as indicated by the fact that fractionation of the polymers with boiling solvents always gives polymeric fractions having practically the same structure of the pristine polymers; however, the 1,4 content seems to be higher for poly(1,3-butadiene)s obtained with catalysts based on vanadium complexes bearing more hindered phosphine ligands ($\theta \geq 150°$).

The influence of the type of alkylating agent (dMAO versus sMAO) on the catalytic activity and selectivity was also investigated: under the same polymerization conditions, catalysts using sMAO as co-catalyst resulted to be less active, giving polymers with higher 1,4 content, the *cis*-1,4 units being predominant (Table 2, entry 4 vs. 5, entry 2 vs. 3, entry 7 vs. 8, and entry 9 vs. 10).

A spread of the weight-average molecular weights, depending on the complex used, ranging from rather low (M_w = 75,000 g/mol, entry 11) to significantly higher (M_w = 296,000 g/mol, entry 16), was observed. The molecular weight distributions (M_w/M_n) were fairly narrows, although tending to a certain broadening in some cases.

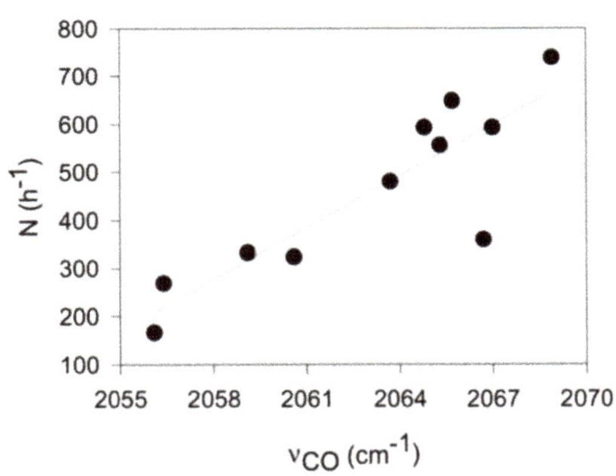

Figure 2. Plot of N (moles of 1,3-butadiene polymerized per mol of vanadium per hour) vs. the phosphine donor ability [as measured by the electronic parameter υ_{CO} based on the carbonyl stretching frequencies in $Ni(CO)_3L$ complexes (L = phosphine ligand)] as defined by Tolman [43,44].

Finally, by comparing the results obtained with the vanadium systems described in the present paper with those obtained with catalysts based on other transition metals phosphine complexes (e.g., Cr, Co, Fe and Ti), we can draw the following conclusions: (i) as far as catalytic activity is concerned, at the same polymerization conditions (i.e., monomer and catalyst concentration, solvent and temperature polymerization), vanadium systems are considerably less active than chromium (e.g., $CrCl_2(dmpe)_2/MAO$) [15] and cobalt (e.g., $CoCl_2(P^iPrPh_2)_2/MAO$) [18] ones, while they are comparable to those titanium-based (e.g., $TiCl_2(dmpe)_2/MAO$) [39] and clearly more active than the iron-based ones [39] just because the latter are practically inactive; (ii) concerning instead the catalytic selectivity, vanadium systems are characterized by the lowest stereospecificity since they essentially provide poly(1,3-butadiene)s with mixed 1,4/1,2 structure, while titanium based systems allow to obtain predominantly *cis*-1,4 poly(1,3-butadiene)s (≥80%) [39] and cobalt based systems can give highly *cis*-1,4 or highly 1,2 syndiotactic poly(1,3-butadiene)s, depending on the type of phosphine ligand on the cobalt atom [18–20,39].

3. Materials and Methods

3.1. General Procedures and Materials

Manipulations of air- and/or moisture-sensitive materials were carried out under an inert atmosphere using a dual vacuum/nitrogen line and standard Schlenk-line techniques. Nitrogen was purified by passage over columns of $CaCl_2$, molecular sieves and BTS catalysts. THF (Aldrich, St. Louis, MO, USA, ≥99.9%) was refluxed over Na/benzophenone alloy for eight h and then distilled and stored over molecular sieves. Toluene (Aldrich, >99.5%) was refluxed over Na for 8 h and then distilled and stored over molecular sieves. Pentane (Aldrich, >99%) was refluxed over Na/K alloy for eight h and then distilled and stored over molecular sieves. Tri-normal-propylphosphine (P^nPr_3) (Strem, Newburyport, MA, USA, 95%), tricyclopentylphosphine ($PCyp_3$) (Strem, >97%), tricyclohexylphosphine (PCy_3) (Strem, 97%), methyldiphenylphosphine ($PMePh_2$) (Strem, 99%), dimethylphenylphosphine (PMe_2Ph) (Strem, 99%), ethyldiphenylphosphine ($PEtPh_2$) (Aldrich, 98%), diethylphenylphosphine (PEt_2Ph) (Aldrich, 96%), di-tert-butylmethylphosphine (P^tBu_2Me) (Strem, >98%), iso-propyldiphenylphosphine (P^iPrPh_2) (Aldrich, 97%), cyclohexyldiphenylphosphine ($PCyPh_2$) (Strem, 99%), dicyclohexylphenylphosphine (PCy_2Ph) (Aldrich, 95%), triphenylphosphine (PPh_3) (Aldrich, ≥95%), tri-tert-butylphosphine (P^tBu_3) (Aldrich, 95%), MAO (Aldrich, 10 wt % solution in toluene) and $VCl_3(THF)_3$ (Aldrich) were used as received. Solid MAO (dMAO) was prepared by removing toluene and unreacted free trimethylaluminum under vacuum from commercially available MAO solution (sMAO). The resulting white powder was heated further to 50 °C under vacuum overnight. 1,3-Butadiene (Air Liquide, Milano, Italy, >99.5% pure) was evaporated from the container prior to each run, dried by passing through a column packed with molecular sieves and condensed into the reactor which had been precooled to −20 °C. Deuterated solvent for NMR measurements ($C_2D_2Cl_4$) (Aldrich, >99.5% atom D), was used as received. For the Elemental Analysis a Perkin Elmer CHN Analyzer 2400 Series II (Perkin Elmer, Milano, Italy) was used.

3.2. Synthesis of the V(III)–Phosphine Complexes

3.2.1. Trichlorobis(triphenylphosphine)vanadium(III) [$VCl_3(PPh_3)_2$, **1a**]

PPh_3 (2.08 g, 10.6 mmol) was added to a suspension of $VCl_3(THF)_3$ (0.99 g, 2.65 mmol, P/V = 4:1) in toluene (15 mL) at room temperature. The solution turned rapidly brown, it was kept under stirring overnight, and then concentrated to half of its volume. Pentane (50 mL) was added dropwise and a light brown precipitate was rapidly formed. The suspension was filtered and the solid residue on the filter washed with pentane (3 × 20 mL), then dried under vacuum to give **1a** as lilac powder (Yield, 83%). Attempts to obtain crystals suitable for the molecular structure determination failed. FTIR (cm^{-1}): 3054 (vw), 1589 (w), 1482 (w), 1436 (m), 1158 (m), 1119 (m), 1093 (m), 991 (m), 748 (mw), 723 (s), 692 (s), 533 (s), 496 (mw). Anal. Calcd. for $C_{36}H_{30}Cl_3P_2V$: C, 63.41; H, 4.43; Cl, 15.60; P, 9.08; V, 7.47. Found: C, 63.30; H, 4.50; Cl, 15.50; P, 9.0; V, 7.60.

3.2.2. Trichlorobis(methyldiphenylphosphine)vanadium(III) [$VCl_3(PMePh_2)_2$, **1b**]

1b was synthesized in the same way as described for **1a** using $PMePh_2$ (2.19 g, 11 mmol) and $VCl_3(THF)_3$ (1.02 g, 2.73 mmol, P/V = 4:1), and it was obtained as red-purple powder (Yield, 90%). The solid was then extracted continuously with boiling pentane. Crystals of **1b** were formed directly on the bottom of the Schlenk tube during the extraction, and further crops of crystals were obtained by cooling the supernatant pentane solution at −30 °C [30]. FTIR (cm^{-1}): 3053 (w), 1588 (w), 1483 (mw), 1435 (m), 1296 (w), 1130 (s), 1096 (m), 1026 (w), 994 (m), 885 (s), 780 (mw) 741 (vs), 690 (vs), 505 (s), 477 (m). Anal. Calcd. for $C_{26}H_{26}Cl_3P_2V$: C, 55.99; H, 4.70; Cl, 19.07; P, 11.11; V, 9.13. Found: C, 56.20; H, 4.60; Cl, 19.20; P, 11.10; V, 9.20.

3.2.3. Trichlorobis(ethyldiphenylphosphine)vanadium(III) [$VCl_3(PEtPh_2)_2$, **1c**]

1c was synthesized in the same way as described for **1a** using $PEtPh_2$ (2.9 g, 13.7 mmol) and $VCl_3(THF)_3$ (1.28 g, 3.42 mmol), and it was obtained as lilac powder (Yield, 91%). The solid was then extracted continuously with boiling pentane. Crystals of **1c** were formed directly on the bottom of the Schlenk tube during the extraction, and further crops of crystals were obtained by cooling the supernatant pentane solution at −30 °C. FTIR (cm^{-1}): 3054 (vw), 2927 (vs), 1587 (vw), 1484 (w), 1436 (m), 1144 (m), 1120 (m), 1098 (m), 1028 (w), 997 (m), 908 (w), 734(s), 691 (vs), 539 (m), 506 (m), 481 (m). Anal. Calcd. for $C_{28}H_{30}Cl_3P_2V$: C, 57.41; H, 5.16; Cl, 18.16; P, 10.58; V, 8.70. Found: C, 57.40; H, 5.10; Cl, 18.20; P, 10.07; V, 8.60.

3.2.4. Trichlorobis(iso-propyldiphenylphosphine)vanadium(III) [$VCl_3(P^iPrPh_2)_2$, **1d**]

1d was synthesized in the same way as described for **1a** using P^iPrPh_2 (2.0 g, 8.76 mmol) and $VCl_3(THF)_3$ (0.82 g, 2.19 mmol), and it was obtained as pink powder (Yield, 91%). FTIR (cm^{-1}): 3054 (vw), 2963 (w), 1587 (w), 1482 (w), 1460 (w), 1437 (m), 1153 (m), 1120 (m), 1100 (m), 1026 (m), 992 (m), 879 (w), 740 (s), 691 (vs), 538 (m), 521 (s), 509 (s), 490 (m), 460 (m). Anal. Calcd. for $C_{30}H_{34}Cl_3P_2V$: C, 58.70; H, 5.58. Found: C, 58.64; H, 5.54.

3.2.5. Trichlorobis(cyclohexyldiphenylphosphine)vanadium(III) [$VCl_3(PCyPh_2)_2$, **1e**]

1e was synthesized in the same way as described for **1a** using $PCyPh_2$ (2.4 g, 9.0 mmol) and $VCl_3(THF)_3$ (0.86 g, 2.3 mmol), and it was obtained as light blue powder (Yield, 56%). Attempts to obtain crystals useful to determine the structure of the complex failed. FTIR (cm^{-1}): 3055 (vw), 2930 (m), 2875 (w), 1589 (w), 1483 (w), 1437 (m), 1152 (m), 1118 (s), 1097 (m), 1028 (w), 999 (s), 917 (w), 888 (w), 851 (w), 822 (w), 741 (s), 724 (s), 690 (vs), 554 (s), 529 (vs), 499 (m), 486 (m), 463 (w). Anal. Calcd. for $C_{36}H_{42}Cl_3P_2V$: C, 62.31; H, 6.10; Cl, 15.33; P, 8.93; V, 7.34. Found: C, 62.40; H, 6.30; Cl, 15.50; P, 9.0; V, 7.20.

3.2.6. Trichlorobis(dimethyiphenylphosphine)vanadium(III) [$VCl_3(PMe_2Ph)_2$, **1f**]

1f was synthesized in the same way as described for **1a** using PMe_2Ph (1.06 g, 7.7 mmol) and $VCl_3(THF)_3$ (0.72 g, 1.92 mmol), and it was obtained as pale pink powder (Yield, 87%). Attempts to obtain crystals useful to determine the structure of the complex failed. FTIR (cm^{-1}): 3053 (w), 1435 (mw), 1140 (mw), 997 (mw), 951 (m), 840 (mw), 745 (vs), 693 (vs), 485 (m). Anal. Calcd. for $C_{16}H_{22}Cl_3P_2V$: C, 44.32; H, 5.11. Found: C, 44.50; H, 5.20.

3.2.7. Trichlorobis(diethylphenylphosphine)vanadium(III) [$VCl_3(PEt_2Ph)_2$, **1g**]

PEt_2Ph (1.0 g, 6.1 mmol) was added dropwise to a suspension of $VCl_3(THF)_3$ (0.85 g, 2.26 mmol, P/V = 2.6:1) in THF (25 mL) at room temperature. The solution turned from dark red to green rapidly and was kept under stirring overnight, then concentrated to half of its volume. Pentane (50 mL) was added dropwise and a beige oily product formed. The mixture was kept under stirring overnight. The solvent was discarded, leaving the oily product behind, washed with pentane (3 × 40 mL) and finally dried in vacuo. A successive addition of a large excess of pentane, and leaving at −30 °C for 3 h produced green chunks of **1g** (Yield, 85%). Crystals of **1g** were formed directly on the bottom of the Schlenk tube during the extraction, and further crops of crystals were obtained by cooling the supernatant pentane solution at −30 °C. FTIR (cm^{-1}): 3054 (vw), 2938 (w), 1589 (w), 1485 (w), 1455 (m), 1438 (s), 1398 (m), 1268 (w), 1248 (w), 1130 (vs), 1107 (vs), 1072 (s), 1049 (s), 1020 (s), 1001 (vs), 934 (w), 898 (w), 762 (s), 746 (s), 725 (s), 678 (s), 515 (s), 477 (m). Anal. Calcd. for $C_{20}H_{30}Cl_3P_2V$: C, 49.05; H, 6.17. Found: C, 49.17; H, 6.23.

3.2.8. Trichlorobis(dicyclohexylphenylphosphine)vanadium(III) [$VCl_3(PCy_2Ph)_2$, **1h**]

1h was synthesized in the same way as described for **1g** using PCy_2Ph (1.0 g, 3.65 mmol) and $VCl_3(THF)_3$ (0.55 g, 1.48 mmol), and it was obtained as light green powder (Yield, 76%). Attempts to obtain crystals useful to determine the structure of the complex failed. FTIR (cm^{-1}): 3054 (w), 2924 (vs), 2851 (s), 1588 (w), 1439 (s), 1328 (w), 1299 (w), 1271 (w), 1177 (w), 1125 (m), 1069 (w), 1003 (s), 905 (m), 850 (m), 747 (s), 696 (s), 530 (m), 493 (m). Anal. Calcd. for $C_{36}H_{54}Cl_3P_2V$: C, 61.24; H, 7.71. Found: C, 61.19; H, 7.68.

3.2.9. Trichlorobis(tricyclopentylphosphine)vanadium(III) [$VCl_3(PCyp_3)_2$, **2a**]

2a was synthesized in the same way as described for **1a** using $PCyp_3$ (2.9 g, 13.7 mmol) and $VCl_3(THF)_3$ (0.92 g, 2.44 mmol), and it was obtained as a pink powder (Yield, 65%). The solid was then extracted continuously with boiling pentane. Crystals of **2a** were formed directly on the bottom of the Schlenk tube during the extraction, and further crops of crystals were obtained by cooling the supernatant pentane solution at −30 °C. FTIR (cm^{-1}): 2949 (vs), 2866 (vs), 1448 (m), 1299 (mw), 1261 (mw), 1231 (mw), 1126 (m), 1053 (m), 1028 (m), 994 (vs), 907 (mw), 806 (s), 508 (m), 487 (m). Anal. Calcd. for $C_{30}H_{54}Cl_3P_2V$: C, 56.83; H, 8.59; Cl, 16.78; P, 9.77; V, 8.03. Found: C, 56.90; H, 8.70; Cl, 16.70; P, 9.80; V, 8.0.

3.2.10. Trichlorobis(tricyclohexylphosphine)vanadium(III) [$VCl_3(PCy_3)_2$, **2b**]

2b was synthesized in the same way as described for **1g** using PCy_3 (2.47 g, 8.82 mmol) and $VCl_3(THF)_3$ (0.83 g, 2.20 mmol), and it was obtained as a light grey powder (Yield, 55%). Attempts to obtain crystals useful to determine the structure of the complex failed. FTIR (cm^{-1}): 3277 (vw), 2927 (vs), 2851 (vs), 1446 (s), 1262 (w), 1107 (m), 1006 (vs), 900 (m), 851 (m), 802 (m), 519 (mw), 467 (mw). Anal. Calcd. for $C_{36}H_{66}Cl_3P_2V$: C, 60.21; H, 9.26; Cl, 14.81; P, 8.63; V, 7.09. Found: C, 60.30; H, 9.20; Cl, 14.70; P 8.70 V, 7.30.

3.2.11. Trichlorobis(tri-normal-propylphosphine)vanadium(III) [$VCl_3(P^nPr_3)_2$, **2c**]

2c was synthesized in the same way as described for **1g** using the ligand P^nPr_3 (1.08 g, 6.72 mmol) and $VCl_3(THF)_3$ (1.0 g, 2.7 mmol), and it was obtained as a beige powder (Yield, 35%). Attempts to obtain crystals useful to determine the structure of the complex failed. FTIR (cm^{-1}): 2093 (mw), 2933 (mw), 2874 (mw), 2903 (mw), 1461 (m), 1403 (m), 1380 (mw), 1243 (vw), 1071 (m), 997 (m), 916 (m), 733 (mw). Anal. Calcd. for $C_{18}H_{42}Cl_3P_2V$: C, 45.25; H, 8.86. Found: C, 45.10; H, 8.78.

3.2.12. Trichlorobis(tri-tert-butylphosphine)vanadium(III) [$VCl_3(P^tBu_3)_2$, **2d**]

2d was synthesized in the same way as described above for **1g** using the ligand P^tBu_3 (1.74 g, 8.64 mmol) and $VCl_3(THF)_3$ (0.46 g, 2.16 mmol), and it was obtained as gray/brown powder (Yield, 31%). FTIR (cm^{-1}): 3029 (vw), 2971 (w), 1592(w), 1472 (m), 1402 (w), 1379 (w), 1261 (w), 1175 (m), 1097 (w), 998 (s), 934 (w), 871 (m), 807 (m), 732 (w), 672 (m), 618 (w), 554 (m), 493 (m), 474 (m). Anal. Calcd. for $C_{24}H_{54}Cl_3P_2V$: C, 51.30; H, 9.69; Cl, 18.93; P, 11.02; V, 9.07. Found: C, 51.50; H, 9.50; Cl, 19.10; P, 11.20; V, 9.30.

3.3. X-ray Crystallographic Studies

The intensity data were collected on a Bruker Smart Apex CCD area detector using graphite-monochromated Mo Kα radiation (λ = 0.71073 Å). Data reduction was made using SAINT programs; absorption corrections based on multiscan were obtained by SADABS [45]. The structures were solved by SHELXS-97 [46], and refined on F^2 by full-matrix least-squares using SHELXL-14 [47]. All the non-hydrogen atoms were refined anisotropically, hydrogen atoms were included as 'riding' and not refined. The isotropic thermal parameters of H atoms were fixed at 1.2 (1.5 for methyl groups) times the equivalent thermal parameter of the atoms to which they are bonded. Crystal data and results of the refinement: (**1c**) air-sensitive orange prism 0.50 × 0.20 × 0.20 mm, Mr = 585.75, monoclinic, space group $P2_1/c$, a = 8.3234(12) Å, b = 9.7489(14) Å, c = 34.500(5) Å, β = 93.804(2)°, V = 2793.3(7) Å^3, Z = 4, T = 130(2) K, μ = 0.773 mm^{-1}. 37,054 measured reflections, 5141 independent reflections, 4377 reflections with I > 2σ(I), 2.36 < 2θ < 50.91°, R_{int} = 0.0468. Refinement on 5141 reflections, 309 parameters. Final R = 0.0261, wR = 0.0564 for data with $F^2 > 2\sigma(F^2)$, S = 1.036, $(\Delta/\sigma)_{max}$ = 0.001, $\Delta\rho_{max}$ = 0.286, $\Delta\rho_{min}$ = −0.249 eÅ^{-3}; (**1g**) air-sensitive light blue tablet 0.25 × 0.12 × 0.10 mm, Mr = 489.67, orthorhombic, space group $P2_12_12_1$, a = 11.9609(5) Å, b = 13.8664(5) Å, c = 13.9617(5) Å, V = 2315.61(15) Å^3, Z = 4, T = 150(2) K, μ = 0.916 mm^{-1}. 46,227 measured reflections, 6991 independent reflections, 6547 reflections with I > 2σ (I), 4.49 < 2θ < 60.84°, R_{int} = 0.0364. Refinement on 6991 reflections, 239 parameters. Final R = 0.0485, wR = 0.1459 for data with $F^2 > 2\sigma(F^2)$, S = 1.076, $(\Delta/\sigma)_{max}$ = 0.001, $\Delta\rho_{max}$ = 1.109, $\Delta\rho_{min}$ = −1.955 eÅ^{-3}; (**2a**) air-sensitive red tablet 0.50 × 0.17 × 0.15 mm, Mr = 633.96, triclinic, space group P-1, a = 9.9362(15) Å, b = 11.7793(18) Å, c = 14.673(2) Å, α = 83.712(2)°, β = 89.645(2)°, γ = 66.242(2)°, V = 1561.0(4) Å^3, Z = 2, T = 100(2) K, μ = 0.696 mm^{-1}. 16,969 measured reflections, 4488 independent reflections, 3859 reflections with I > 2σ(I), 2.80 < 2θ < 46.60°, R_{int} = 0.0395. Refinement on 4488 reflections, 325 parameters. Final R = 0.0385, wR = 0.0911 for data with $F^2 > 2\sigma(F^2)$, S = 1.066, $(\Delta/\sigma)_{max}$ = 0.002, $\Delta\rho_{max}$ = 0.606, $\Delta\rho_{min}$ = −0.940 eÅ^{-3}. Table S1 contains cartesian coordinates of the modelled compounds at their UM06/6-311G(d) optimized geometry; while CCDC 1583626–1583628 contain supplementary crystallographic data which can be obtained free of charge from The Cambridge Crystallographic Data Centre via www.ccdc.cam.ac.uk/data_request/cif.

3.4. Computational Details

Geometry optimization of the isolated, gas phase vanadium(III) complexes were performed with the 6-311g(d) basis set by the Gaussian16 program package [48], starting from the respective X-ray molecular structures. For comparative purposes, two exchange-correlation functionals have been used: the M06 functional [34], due to its specific parametrization on organometallic complexes, and the B3LYP functional [35–37], the most used functional for geometry optimizations. According to previous magnetic measurements on **1b** [25], complexes were modeled in their triplet state through the unrestricted formalism.

3.5. Polymerization of 1,3-Butadiene

Polymerizations were carried out in a 25 mL round-bottomed Schlenk flask. Prior to starting polymerization, the reactor was heated to 110 °C under vacuum for one h and backfilled with nitrogen. 1,3-Butadiene was condensed into the Schlenk flask kept at −20 °C, then toluene was added and the solution was brought to the desired polymerization temperature. The aluminum alkyl and a toluene solution (2 mg/mL) of the vanadium complex were then added in that order. Polymerization was stopped with methanol containing a small amount of hydrochloric acid. The polymer obtained was then coagulated by adding 40 mL of a methanol solution containing 4% of Irganox® 1076 antioxidant (BASF, Ludwigshafen am Rhein, Germany), repeatedly washed with fresh methanol and finally dried in vacuum at room temperature to constant weight.

3.6. Polymer Characterization

For all the mentioned catalysts the FTIR spectrum was collected. Attenuated total reflectance (ATR)-Fourier transform infrared spectroscopy (FTIR) spectra were recorded at room temperature in the 600–4000 cm^{-1} range with a resolution of 4 cm^{-1} using a Perkin Elmer Spectrum Two spectrometer (Perkin Elmer, Milano, Italy.). NMR spectra were recorded on a Bruker NMR advance 400 Spectrometer (Bruker Italia Srl, Milano, Italy) operating at 400 MHz (1H) and 100.58 MHz (^{13}C) working in the PFT mode at 103 °C . NMR samples were prepared dissolving from 60 to 80 mg of polymer in about three mL of $C_2D_2Cl_4$ in 10 mm probes and referred to hexamethyldisiloxane (HMDS), as internal standard. The relaxation delay was 16 s. The molecular weight average (M_w) and the molecular weight distribution (M_w/M_n) were obtained by a high temperature Waters GPCV2000 size exclusion chromatography (SEC) system equipped with a refractometer detector. The experimental conditions consisted of three PL Gel Olexis columns, ortho-dichlorobenzene (DCB) as the mobile phase, 0.8 mL/min flow rate, and 145 °C temperature. The calibration of the SEC system was constructed using eighteen narrow M_w/M_n PS standards with molar weights ranging from 162 to 5.6×10^6 g/mol. For SEC analysis, about 12 mg of polymer was dissolved in five mL of DCB with 0.05% of BHT as antioxidant.

4. Conclusions

A series of V(III) complexes with aryl and alkyl phosphines were synthesized and characterized, including single crystal X-ray molecular structure determination of $VCl_3(PEtPh_2)_2$, $VCl_3(PEt_2Ph)_2$ and $VCl_3(PCyp_3)_2$. The complexes were then successfully used, in combination with various aluminoxanes, for the polymerization of 1,3-butadiene. In general, we obtained polymers having different microstructures, depending on the phosphine ligand and the Al-based activator. Catalysts based on vanadium complexes with more hindered ligands gave poly(1,3-butadiene)s with higher 1,4 (*cis* + *trans*) content ($\geq$80 mol %); catalysts with sMAO as co-catalyst were found to give higher *cis*-1,4 content ($\geq$65 mol %). The type of ligand and co-catalyst were also found to have some influence on the catalyst activity. In general catalysts based on complexes with aryl phosphines exhibited higher activity than those bearing alkyl phosphines; catalysts using dMAO instead of sMAO as co-catalyst resulted to be much more active, allowing to reach complete monomer conversion, thus indicating that vanadium catalysts in case of conjugated diolefins polymerization are not subject to deactivation and do not need the use of a reoxidant, as it happens in case of monoolefins.

Supplementary Materials: The following are available online at http://www.mdpi.com/2073-4344/7/12/369/s1. Table S1: Cartesian coordinates of the modeled compounds at their UM06/6-311G(d) optimized geometry, Figure S1: ^{1}H NMR spectrum of poly(1,3-butadiene) (Table 2, entry 4), Figure S2: ^{13}C NMR spectrum of poly(1,3-butadiene) (Table 2, entry 4), Figure S3: ^{1}H NMR spectrum of poly(1,3-butadiene) (Table 2, entry 14), Figure S4: ^{13}C NMR spectrum of poly(1,3-butadiene) (Table 2, entry 14).

Author Contributions: G.L. performed the experiments, analyzed the data, and contributed to write the paper. G.Z. performed the experiments and a part of them were reproduced by I.P. and A.S.; A.F. determined the X-ray molecular structures of the complexes, performed the quantum-mechanical calculations and contributed to write the paper. G.R. was the principal investigator, conceived and designed the experiments, analyzed the data and wrote the paper.

Conflicts of Interest: The authors declare no conflict of interest.

References

1. Porri, L.; Giarrusso, A. Part II. In *Conjugated Diene Polymerization in Comprehensive Polymer Science*; Pergamon: Oxford, UK, 1989; Volume 4, pp. 53–108.
2. Thiele, S.K.-H.; Wilson, D.R. Alternate Transition Metal Complex Based Diene Polymerization. *J. Macromol. Sci. Part C Polym. Rev.* **2003**, *C43*, 581–628. [CrossRef]
3. Friebe, L.; Nuyken, O.; Obrecht, W. Neodymium-Based Ziegler/Natta Catalysts and their Application in Diene Polymerization. *Adv. Polym. Sci.* **2006**, *204*, 1–154.

4. Natta, G.; Porri, L.; Corradini, P.; Morero, D. Polimerizzazioni stereospecifiche di diolefine coniugate. *Chim. Ind.* **1958**, *40*, 362–371.
5. Natta, G.; Porri, L.; Carbonaro, A. Stereospecificity of homogeneous catalysts prepared from vanadium trichloride in the polymerization of conjugated diolefins. *Atti Accad. Naz. Lincei Cl. Sci. Fis. Mat. Nat. Rend.* **1961**, *31*, 189; reprinted in *Chem. Abstr.* **1962**, *57*, 4848.
6. Porri, L.; Carbonaro, A.; Ciampelli, F. Copolymerization of 1,3-butadiene and 1,3-pentadiene with homogeneous Al (C2H5)2Cl-vanadium compounds catalyst systems. I. Preparation and properties of the copolymers. *Makromol. Chem.* **1963**, *61*, 90–103. [CrossRef]
7. Ricci, G.; Italia, S.; Comitani, C.; Porri, L. Polymerization of conjugated dialkenes with transition metal catalysts. Influence of methylaluminoxane on catalyst activity and stereospecificity. *Polym. Commun.* **1991**, *32*, 514–517.
8. Ricci, G.; Zetta, L.; Alberti, E.; Motta, T.; Canetti, M.; Bertini, F. Butadiene–isoprene copolymerization with V(acac)$_3$-MAO. Crystalline and amorphous *trans*-1,4 copolymers. *J. Polym. Sci. Part A Polym. Chem.* **2007**, *45*, 4635–4646. [CrossRef]
9. Colamarco, E.; Milione, S.; Cuomo, C.; Grassi, A. Homo- and Copolymerization of Butadiene Catalyzed by an Bis(imino)pyridyl Vanadium Complex. *Macromol. Rapid Commun.* **2004**, *25*, 450–454. [CrossRef]
10. Natta, G.; Porri, L.; Zanini, G.; Fiore, L. Stereospecific polymerization of conjugated diolefins. IV. Preparation of syndiotactic 1,2-polybutadiene. *Chim. Ind. (Milan)* **1959**, *41*, 526–532; reprinted in *Chem. Abstr.* **1960**, *54*, 1258.
11. Porri, L.; Ricci, G.; Giarrusso, A. *Metalorganic Catalysts for Synthesis and Polymerization*; Kaminsky, W., Ed.; Springer: Berlin/Heidelberg, German, 1999; p. 519.
12. Porri, L.; Giarrusso, A.; Ricci, G. *Metallocene-Based Polyolefins*; Scheirs, J., Kaminsky, W., Eds.; John Wiley & Sons Ltd.: London, UK, 2000; p. 115.
13. Ricci, G.; Panagia, A.; Porri, L. Polymerization of 1,3-dienes with catalysts based on mono-and bis-cyclopentadienyl derivatives of vanadium. *Polymer* **1996**, *37*, 363–365. [CrossRef]
14. Bradley, S.; Camm, K.D.; Furtado, S.J.; Gott, A.L.; McGowan, P.C.; Podesta, T.J.; Thornton-Pett, M. Synthesis and Structure of Amino-Functionalized Cyclopentadienyl Vanadium Complexes and Evaluation of Their Butadiene Polymerization Behavior. *Organometallics* **2002**, *21*, 3443–3453. [CrossRef]
15. Ricci, G.; Battistella, M.; Porri, L. Chemoselectivity and Stereospecificity of Chromium(II) Catalysts for 1,3-Diene Polymerization. *Macromolecules* **2001**, *34*, 5766–5769. [CrossRef]
16. Ricci, G.; Forni, A.; Boglia, A.; Sonzogni, M. New Chromium(II) Bidentate Phosphine Complexes: Synthesis, Characterization, and Behavior in the Polymerization of 1,3-Butadiene. *Organometallics* **2004**, *23*, 3727–3732. [CrossRef]
17. Ricci, G.; Boglia, A.; Motta, T. Synthesis of new Cr(II) complexes with bidentate phosphine ligands and their behavior in the polymerization of butadiene. Influence of the phosphine bite angle on catalyst activity and stereoselectivity. *J. Mol. Catal. A Chem.* **2007**, *267*, 102–107. [CrossRef]
18. Ricci, G.; Forni, A.; Boglia, A.; Motta, T.; Zannoni, G.; Canetti, M.; Bertini, F. Synthesis and X-ray Structure of $CoCl_2(P^iPrPh_2)_2$. A New Highly Active and Stereospecific Catalyst for 1,2 Polymerization of Conjugated Dienes When Used in Association with MAO. *Macromolecules* **2005**, *38*, 1064–1070. [CrossRef]
19. Ricci, G.; Forni, A.; Boglia, A.; Motta, T. Synthesis, structure, and butadiene polymerization behavior of alkylphosphine cobalt(II) complexes. *J. Mol. Catal. A Chem.* **2005**, *226*, 235–241. [CrossRef]
20. Ricci, G.; Forni, A.; Boglia, A.; Sommazzi, A.; Masi, F. Synthesis, structure and butadiene polymerization behavior of $CoCl_2(PR_xPh_{3-x})_2$ (R = methyl, ethyl, propyl, allyl, isopropyl, cyclohexyl; x = 1, 2). Influence of the phosphorous ligand on polymerization stereoselectivity. *J. Organomet. Chem.* **2005**, *690*, 1845–1854. [CrossRef]
21. Ricci, G.; Motta, T.; Boglia, A.; Alberti, E.; Zetta, L.; Bertini, F.; Arosio, P.; Famulari, A.; Meille, S.V. Synthesis, Characterization, and Crystalline Structure of Syndiotactic 1,2-Polypentadiene: The Trans Polymer. *Macromolecules* **2005**, *38*, 8345–8352. [CrossRef]
22. Ricci, G.; Boglia, A.; Motta, T.; Bertini, F.; Boccia, A.C.; Zetta, L.; Alberti, E.; Famulari, A.; Arosio, P.; Meille, S.V. Synthesis and structural characterization of syndiotactic *trans*-1,2 and *cis*-1,2 polyhexadienes *J. Polym. Sci. Part A Polym. Chem.* **2007**, *45*, 5339–5353. [CrossRef]
23. Ricci, G.; Leone, G.; Boglia, A.; Boccia, A.C.; Zetta, L. *cis*-1,4-*alt*-3,4 Polyisoprene: Synthesis and Characterization. *Macromolecules* **2009**, *42*, 9263–9267. [CrossRef]

24. Boccia, A.C.; Leone, G.; Boglia, A.; Ricci, G. Novel stereoregular *cis*-1,4 and *trans*-1,2 poly(diene)s: Synthesis, characterization, and mechanistic considerations. *Polymer* **2013**, *54*, 3492–3503. [CrossRef]
25. Bansemer, R.L.; Huffman, J.C.; Caulton, K.G. Synthesis and characterization of $VCl_3(PMePh_2)_2$. *Inorg. Chem.* **1985**, *24*, 3003–3006. [CrossRef]
26. Holt, D.G.L.; Larkworthy, L.F.; Povey, D.C.; Smith, G.W.; Leigh, G.J. Synthesis and structural studies of complexes of vanadium(II) and vanadium(III) halides with tertiary phosphines. *Inorg. Chim. Acta* **1993**, *207*, 11–19. [CrossRef]
27. Bultitude, J.; Larkworthy, L.F.; Povey, D.C.; Smith, G.W.; Dilworth, J.R. The preparation and crystal and molecular structures of trichlorobis(methyldiphenylphosphine)vanadium(III) and its acetonitrile adduct. *J. Chem. Soc. Dalton Trans.* **1986**, 2253–2258. [CrossRef]
28. Nieman, J.; Teuben, J.H.; Huffman, J.C.; Caulton, K.G. Preparation and characterization of mono-cyclopentadienylvanadium dihalide bis-phosphine complexes—Crystal-structure of (eta-5-C_5H_5) $VCl_2(PMe_3)_2$. *J. Org. Chem.* **1983**, *255*, 193–204. [CrossRef]
29. Issleib, V.K.; Bohn, V. Phosphin- und Phosphinoxydkomplexe des 3wertigen Vanadins. *Z. Anorg. Allg. Chem.* **1959**, *301*, 188–196. [CrossRef]
30. Leone, G.; Pierro, I.; Zanchin, G.; Forni, A.; Bertini, F.; Rapallo, A.; Ricci, G. Vanadium(III)–catalyzed copolymerization of ethylene with norbornene: Microstructure at tetrad level and reactivity ratios. *J. Mol. Catal. A Chem.* **2016**, *424*, 220–231. [CrossRef]
31. Ricci, G.; Leone, G.; Sommazzi, A.; Forni, A.; Masi, F. Phosphinic Vanadium Complex, Catalytic System Comprising Said Phosphinic Vanadium Complex and Process for the (co)polymerization of Conjugated Dienes. U.S. Patent Application 20,170,275,312, 18 August 2017.
32. Cotton, F.A.; Lu, J. EPR and Crystallographic Studies of Some Reaction Products of VCl_4, $NbCl_4$, and $TaCl_4$ with Trialkyl- and Triarylphosphines. *Inorg. Chem.* **1995**, *34*, 2639–2644. [CrossRef]
33. Ernst, R.D.; Freeman, J.W.; Stahl, L.; Wilson, D.R.; Arif, A.M.; Nuber, B.; Ziegler, M.L. Longer but Stronger Bonds: Structures of PF_3, $P(OEt)_3$, and PMe_3 Adducts of an Open Titanocene. *J. Am. Chem. Soc.* **1995**, *117*, 5075–5081.
34. Zhao, Y.; Truhlar, D.G. The M06 suite of density functionals for main group thermochemistry, thermochemical kinetics, noncovalent interactions, excited states, and transition elements: Two new functionals and systematic testing of four M06-class functionals and 12 other functionals. *Theor. Chem. Acc.* **2008**, *120*, 215–241.
35. Lee, C.; Yang, W.; Parr, R.G. Development of the Colle-Salvetti correlation-energy formula into a functional of the electron density. *Phys. Rev. B* **1988**, *37*, 785–789. [CrossRef]
36. Becke, A.D. Density-functional thermochemistry. III. The role of exact exchange. *J. Chem. Phys.* **1993**, *98*, 5648–5652. [CrossRef]
37. Vosko, S.H.; Wilk, L.; Nusair, M. Accurate spin-dependent electron liquid correlation energies for local spin density calculations: A critical analysis. *Can. J. Phys.* **1980**, *58*, 1200–1211. [CrossRef]
38. Aakeröy, C.B.; Evans, T.A.; Seddon, K.R.; Pálinkó, I. The C–H···Cl hydrogen bond: Does it exist? *New J. Chem.* **1999**, *23*, 145–152. [CrossRef]
39. Ricci, G.; Sommazzi, A.; Masi, F.; Ricci, M.; Boglia, A.; Leone, G. Well Defined Transition Metal Complexes with Phosphorus and Nitrogen Ligands for 1,3-Dienes Polymerization. *Coord. Chem. Rev.* **2010**, *254*, 661–676. [CrossRef]
40. Christman, D.L. Preparation of polyethylene in solution. *J. Polym. Sci. Part A1* **1972**, *10*, 471–487. [CrossRef]
41. Gumboldt, V.A.; Helberg, J.; Schleitzer, G. Makromol. Chem. Über die reaktivierung der bei der äuthylen/propylen-copolymerisation verwendeten vanadium-katalysatoren. *Macromol. Chem. Phys.* **1967**, *101*, 229–245. [CrossRef]
42. Tait, P.J.T.; Watkins, N.D. *Monoalkene Polymerization: Mechanisms in: Comprehensive Polymer Science*; Eastmond, G.C., Ledwith, A., Russo, S., Sigwalt, P., Eds.; Pergamon Press Ltd.: Oxford, UK, 1989; Volume 4, Part II; p. 27.
43. Tolman, C.A. Electron donor-acceptor properties of phosphorus ligands. Substituent additivity. *J. Am. Chem. Soc.* **1970**, *92*, 2953–2956. [CrossRef]
44. Tolman, C.A. Steric effects of phosphorus ligands in organometallic chemistry and homogeneous catalysis. *Chem. Rev.* **1977**, *77*, 313–348. [CrossRef]
45. Bruker. *SMART, SAINT and SADABS*; Bruker AXS Inc.: Madison, WI, USA, 1997.
46. Sheldrick, G.M. A short history of SHELX. *Acta Cryst.* **2008**, *A64*, 112–122. [CrossRef] [PubMed]

47. Sheldrick, G.M. Crystal structure refinement with SHELXL. *Acta Cryst.* **2015**, *C71*, 3–8.
48. Frisch, M.J.; Trucks, G.W.; Schlegel, H.B.; Scuseria, G.E.; Robb, M.A.; Cheeseman, J.R.; Montgomery, J.A., Jr.; Vreven, T.; Kudin, K.N.; Burant, J.C.; et al. *Gaussian 16, Revision A.03*; Gaussian, Inc.: Wallingford, CT, USA, 2016.

catalysts

MDPI

Article

Stereoselective Copolymerization of Styrene with Terpenes Catalyzed by an *Ansa*-Lanthanidocene Catalyst: Access to New Syndiotactic Polystyrene-Based Materials

Eva Laur [1], Alexandre Welle [2], Aurélien Vantomme [2], Jean-Michel Brusson [3], Jean-François Carpentier [1,*] and Evgueni Kirillov [1,*]

1. Institut des Sciences Chimiques de Rennes, CNRS, Université de Rennes, UMR 6226, F-35042 Rennes, France; eva.laur@univ-rennes1.fr
2. Total Raffinage Chimie Research, Zone Industrielle C, B-7181 Feluy, Belgium; alexandre.welle@univ-rennes1.fr (A.W.); Aurelien.vantomme@univ-rennes1.fr (A.V.)
3. Total S.A., Direction scientifique, 24 Cours Michelet, F-92069 Paris La Défense CEDEX, France; jean-Michel.Brusson@univ-rennes1.fr

* Correspondence: jean-francois.carpentier@univ-rennes1.fr (J.-F.C.); evgueni.kirillov@univ-rennes1.fr (E.K.); Tel.: +33-223-235-950 (J.-F.C.); +33-223-236-118 (E.K.)

Received: 3 November 2017; Accepted: 25 November 2017; Published: 27 November 2017

Abstract: The copolymerization of bio-renewable β-myrcene or β-farnesene with styrene was examined using an *ansa*-neodymocene catalyst, affording two series of copolymers with high styrene content and unprecedented syndioregularity of the polystyrene sequences. The incorporation of terpene in the copolymers ranged from 5.6 to 30.8 mol % (β-myrcene) and from 2.5 to 9.8 mol % (β-farnesene), respectively. NMR spectroscopy and DSC analyses suggested that the microstructure of the copolymers consists of 1,4- and 3,4-poly(terpene) units randomly distributed along syndiotactic polystyrene chains. The thermal properties of the copolymers are strongly dependent on the terpene content, which is easily controlled by the initial feed. The terpolymerization of styrene with β-myrcene in the presence of ethylene was also examined.

Keywords: copolymerization; rare-earth catalyst; styrene; β-myrcene; farnesene; ethylene; bio-sourced monomers; stereocontrol; syndiotactic

1. Introduction

The sustainability and possible eco-compatibility of bio-based polymers has raised much interest in finding inexpensive alternatives to petroleum-sourced materials. Among the wide variety of renewable monomers, terpenes are particularly interesting because of their abundance in nature, and their reactive conjugated 1,3-diene framework. For instance, β-myrcene, or to a lesser extent β-farnesene, can polymerize via anionic or radical polymerization to form elastomeric materials [1]. In a few rare cases, the coordination (catalytic) polymerization of such monomers has been reported [2–6]. On the other hand, syndiotactic polystyrene (sPS) is a very attractive material due to its high melting point (ca. 270 °C), high crystallinity, low dielectric constant and good chemical and heat resistances [7,8]. However, its processability is limited by its high melting point and its brittleness. Syndioselective copolymerization of styrene with small amounts of a second monomer is one of the most common ways to tune the properties of sPS, and thus to improve its processability [9–11]. A few examples of styrene-myrcene polymers have already been reported, mostly via anionic or radical copolymerizations [12–14]. To our knowledge, only one paper has addressed the coordination copolymerization of styrene with β-myrcene (My) in the presence of the ternary system

$Cp^*La(BH_4)_2(THF)_2$/*n*BuEtMg/Al*i*Bu_3, affording poly(S-*co*-1,4-*trans*-My) containing high amounts of myrcene (66–96 mol %) [15]. Yet, nothing has been reported about styrene-myrcene copolymers containing stereoregular PS sequences, nor about the copolymerization of styrene with β-farnesene.

Very recently, we reported new neutral allyl {Cp/Flu}-type neodymium and samarium complexes of the type {$R_2C(C_5H_4)$(R′R′Flu)}Ln(1,3-$C_3H_3(SiMe_3)_2$)$(THF)_x$ (Flu = 9-fluorenyl) that act as single-component catalysts for homo- and copolymerization of styrene with ethylene, featuring high syndioselectivity, good productivities and good control over the incorporated ethylene content [16]. Previous work conducted by our group also demonstrated syndioselective styrene/isoprene (/ethylene) co(ter)polymerizations using the neutral {(Me_2C(Cp)(Flu)}Nd(C_3H_5)(THF) complex [17]. In this paper, we report on the copolymerization of styrene with β-myrcene or farnesene with such catalyst systems. The synthesis of new poly(S-*co*-My) and poly(S-*co*-Fa) copolymers with high styrene content and containing syndiotactic polystyrene sequences is described. Styrene/ethylene/β-myrcene terpolymerizations were also attempted.

2. Results and Discussion

2.1. Homopolymerization of Myrcene and Farnesene

Homopolymerization of β-myrcene with {$Me_2C(C_5H_4)$(2,7-*t*Bu_2Flu)}Nd(1,3-$C_3H_3(SiMe_3)_2$)(THF) (**1**) in the presence of 10 equiv of $(nBu)_2Mg$ as scavenger was first assessed (Table 1, entries 1 and 2) [16]; this showed, however, very low productivity (0.74 kg·mol^{-1}·h^{-1} at 60 °C; 5.0 kg·mol^{-1}·h^{-1} at 120 °C). Signals in the ^{1}H NMR spectrum of the recovered poly(My) recorded in 1,1,2,2-tetrachloroethane-d_2 were assigned using the previously reported NMR data for poly(β-myrcene) samples [4,18]. Two distinct signals at δ 4.82 and 5.19 ppm revealed the presence of both 1,4- and 3,4-sequences (Figure 1). A 50:50 ratio of 1,4-/3,4- was obtained at T_{polym} = 60 °C; an increase of the polymerization temperature induced an increase in 3,4-insertions (1,4-/3,4- = 25:75 for T_{polym} = 120 °C, see the Supporting Information, Figure S1). The ^{13}C{^{1}H} NMR spectrum also showed characteristic signals corresponding to both 1,4- and 3,4-regioinsertions (Supporting Information, Figures S2 and S3); however, the resolution was not sufficient to distinguish unambiguously the signals corresponding to *cis*- and *trans*-1,4 insertions. The molecular weight of the poly(My) was determined by GPC (gel permeation chromatography) and found to be relatively low (M_n = 3900 g·mol^{-1}). DSC (differential scanning calorimetry) analyses revealed a T_g of ca. −58 °C, consistent with a 50:50 ratio of 1,4- and 3,4-segments [4,15]. Farnesene homopolymerization was also explored (Table 1, entries 11 and 12); however, the productivities were even lower than for β-myrcene homopolymerization (0.28–0.34 kg·mol^{-1}·h^{-1}) and a lower monomer loading was required to recover some poly(Fa) (i.e., $[Fa]_0/[Nd]_0$ = 200 instead of $[My]_0/[Nd]_0$ = 1000). NMR data recorded for the recovered polymer were consistent with those reported in the literature [6,19], and revealed a poly(Fa) microstructure made of a mixture of 1,4- and 3,4-insertions (Supporting Information, Figures S6 and S7). Note that some of those poly(Fa) materials were found to be hardly or even not at all soluble in most of the common solvents (CH_2Cl_2, $CHCl_3$, THF, 1,2,4-trichlorobenzene), possibly due to the presence of reticulated materials. Thus, DSC and GPC analyses were most often uninformative.

Table 1. Styrene/β-myrcene and styrene/farnesene copolymerizations catalyzed by **1** [a].

Entry	Comonomer	$[St]_0$ [M]	$[St]_0/[Nd]$	$[Comon]_0/[St]_0$	T_{polym} [°C]	Time [min]	Prod. [b] [kg·mol^{-1}·h^{-1}]	Conv. St [%]	Conv. Comon [%]	Comon. Inc. [c] [mol %]	T_m [d] [°C]	T_c [d] [°C]	T_g [d] [°C]	ΔH_m [d] [J·g^{-1}]	$M_n \times 10^{-3}$ [g·mol^{-1}] [e]	$Đ_M$ [e]
1 [f]	My	-	0	∞	60	1320	0.74	-	11.6	100	*no*	*no*	−58	*no*	3.87	1.33
2 [f]	My	-	0	∞	120	1320	5.0	-	78.8	100	*no*	*no*	−44	*no*	5.43	1.53
3	-	4.4	8000	0	60	120	248	60	-	0	251,265	231	100	30.0	22.6 [k]	2.5 [k]
4	My	4.4	8000	0.125	60	120	127	28.2	13.1	5.6	208,226	176	73	14.4	23.2	1.8
5 [g]	My	4.4	16,000	0.125	60	120	114	12.6	6.2	5.7	220,235	187	78	20.4	26.8	1.4
6	My	4.4	8000	0.25	60	120	56	11.6	5.7	10.9	206	156	70	9.9	25.1	1.3
7 [h]	My	4.4	20,000	0.25	120	120	374	27.1	26.7	19.7 [g]	no	no	45	no	30.3	1.8
8	My	4.4	8000	0.5	60	180	20	5.4	2.4	18.1	*no*	*no*	48	*no*	36.6	2.7
9 [i]	My	4.4	4000	0.5	60	180	54	29.2	14.8	20.0	*no*	*no*	33	*no*	28.2	2.3
10	My	3.3	6000	1	60	360	6.4	3.9	1.7	30.8	*no*	*no*	12	*no*	35.3	2.3
11 [j]	Fa	-	0	∞	60	1320	0.28	-	15.1	100	*nd*	*nd*	*nd*	*nd*	*nd*	*nd*
12 [j]	Fa	-	0	∞	120	1320	0.34	-	18.6	100	*nd*	v	*nd*	*nd*	*nd*	*nd*
13	Fa	4.4	8000	0.125	60	120	198	45.1	9.4	2.5	246 257	214	83	29.2	31.4 [k]	3.9 [k]
14	Fa	4.4	8000	0.125	120	240	234	>99	30.0	3.4	227	165	86	23.1	18.4 [k]	2.2 [k]
15	Fa	4.4	8000	0.25	60	120	84	19.2	1.6	2.0	243 257	224	85	31.1	20.8 [k]	3.6 [k,*]
16	Fa	3.8	7000	0.5	60	180	31	11.5	1.1	4.7	225 241	204	66	18.4	14.7 [k]	2.3 [k]
17	Fa	2.7	5000	1	60	360	28	26.6	2.9	9.8	212	149	63	9.0	14.4 [k]	2.9 [k]

[a] General conditions unless otherwise stated: 10 µmol of **1**; solvent = cyclohexane (T_{polym} = 60 °C) or *n*-dodecane (T_{polym} = 120 °C); [Nd] = 5.4 × 10^{-4} mol·L^{-1}; [Mg(*n*Bu)$_2$]/[Nd] = 10; styrene, β-myrcene and farnesene (75:25 mixture of α- and β-isomers) purified through neutral alumina, stirring on CaH_2, trap-to-trap vacuum distillation and stored at −27 °C on 3 Å molecular sieves; *no*: not observed; [b] Productivity calculated over the whole reaction time; [c] Determined by ^{1}H NMR spectroscopy; [d] Determined by DSC from second run; [e] Determined by GPC at 30 °C in THF; [f] β-Myrcene homopolymerization, $[My]_0/[Nd]_0$ = 1000, [Nd] = 3.6 × 10^{-3} mol·L^{-1}, $[My]_0$ = 3.6 mol·L^{-1}; [g] [Nd] = 2.7 × 10^{-4} mol·L^{-1}; [h] [Nd] = 2.2 × 10^{-4} mol·L^{-1}; [i] [Nd] = 1.1 × 10^{-3} mol·L^{-1}; [j] Farnesene homopolymerization, $[Fa]_0/[Nd]_0$ = 200, [Nd] = 1.6 × 10^{-2} mol·L^{-1}, $[Fa]_0$ = 3.2 mol·L^{-1}; [k] Determined by GPC at 135 °C in 1,2,4-trichlorobenzene; * Bimodal distribution.

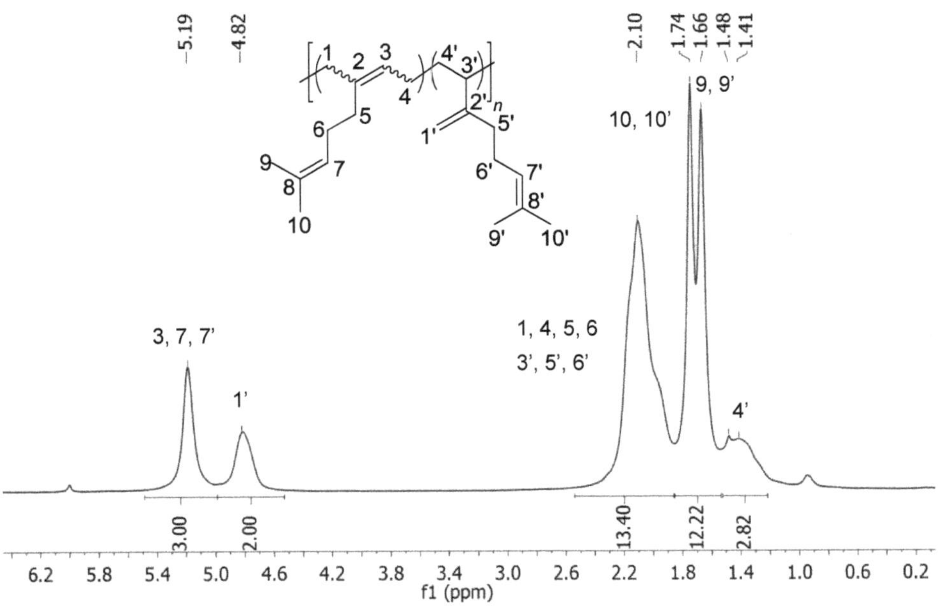

Figure 1. ^{1}H NMR spectrum (500 MHz, 1,1,2,2-tetrachloroethane-d_2, 60 °C) of a poly(β-myrcene) prepared with **1** (Table 1, entry 1).

2.2. Copolymerizations of Styrene with β-Myrcene and Farnesene

A series of styrene/β-myrcene and styrene/farnesene copolymerizations catalyzed by **1** were performed (Scheme 1, Table 1).

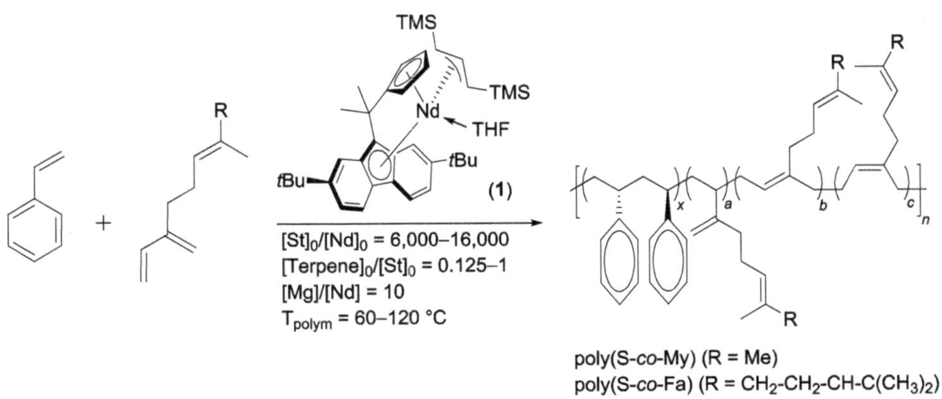

Scheme 1. Styrene/β-myrcene and styrene/farnesene copolymerizations catalyzed by complex **1** (Table 1, entries 4–10 and entries 13–17).

As judged by the relatively narrow monomodal molecular weight distributions ($Đ_M$ = 1.3–2.7), true copolymers were produced. Depending on the quantity of incorporated β-myrcene, the resulting copolymers were either white powders or transparent hard elastomeric materials. On the other hand, all poly(S-*co*-Fa) samples were white powders. The catalyst productivity decreased with an increase of the [terpene]$_0$/[St]$_0$ feed ratio (from 127 to 6.4 kg·mol^{-1}·h^{-1} and from 198 to 28 kg·mol^{-1}·h^{-1} in the cases of styrene/β-myrcene and styrene/farnesene copolymerization, respectively, entries 4–10 and entries 13–17); this is consistent with the fact that **1** is poorly active towards β-myrcene and farnesene homopolymerization (entries 1–2 and 11–12). On the other hand, as the $[\text{terpene}]_0/[\text{St}]_0$ feed ratio was raised, the terpene content in the resulting copolymers increased as expected; thus, a wide range of compositions was obtained (My content = 5.6–30.8 mol % and Fa content = 1.5–9.8 mol %). It should be noted that, under similar copolymerization conditions, the incorporation of farnesene in the copolymer is 2–3 times lower than the incorporation of β-myrcene (for instance, compare entries 4 and 13 or 10 and 17). This difference is due to the fact that commercially available farnesene is a mixture of α- and β-isomers (ca. 75:25 mixture of α/β; see Figure S21); however, only the β-isomer was incorporated, as evidenced from the ^{1}H NMR data (see the Supporting Information, Figure S12), whereas the initial ratio $[\text{Fa}]_0/[\text{St}]_0$ was calculated considering the entire amount of farnesene (i.e., both α- and β-isomers). We assume that the trisubstituted C^3=C bond in the α-isomer, as compared to the disubstituted C^3=C one in the β-isomer, most likely accounts for lower reactivity of the α-isomer (selective polymerization of β-farnesene was already reported in the literature; see [6]). In addition, an increase in polymerization temperature logically induced an increase in productivity (6 times higher when going from 60 °C to 120 °C in the case of styrene/β-myrcene copolymerization), but also an increase in the amount of terpene incorporated in the copolymer (from 10.9 to 19.7 mol % and from 2.0 to 3.4 mol % for β-myrcene and β-farnesene, respectively; entries 6 and 7 and entries 13–14, respectively).

The reactivity ratios of r_{ST} = 2.06 and r_{MY} = 0.37 were determined according to the Fineman−Ross equation (Supporting Information, Figure S8) [20]. These values indicate a preference for the insertion of styrene, regardless of the last inserted monomer unit, consistent with the fact that **1** is 3 orders of magnitude more productive in catalyzing styrene homopolymerization than in catalyzing β-myrcene homopolymerization (entries 1 and 3) [21].

Microstructures of the copolymers were examined by ^{13}C{^{1}H} NMR spectroscopy (Figure 2). Single and relatively sharp signals in the *ipso* carbon (δ 145.2 ppm) and methylene (δ 43.8 ppm) regions of the spectra are indicative of the presence of poly(S-*co*-My) and poly(S-*co*-Fa) with sPS sequences. These signals broadened with an increase of terpene content, due to the increasing presence of styrene-terpene junctions (see Figure 3 for poly(S-*co*-My) copolymers and Supporting Information, Figure S14, for poly(S-*co*-Fa) copolymers). Note that the spectra were recorded using a regular NMR sequence that does not necessarily return quantitative signals; thus, the syndiotacticity content was not quantified. In addition, the presence of both 1,4- and 3,4-units for both copolymers confirms that complex **1** is not regioselective towards β-myrcene or β-farnesene. The non-regioselectivity of **1** in those St/My and St/Fa copolymerizations contrasts with styrene/isoprene copolymers containing regular *trans*-1,4-polyisoprene units obtained in the presence of {CpCMe$_2$Flu}Nd(C$_3$H$_5$)(THF) [17]. We assume that this arises from the more sterically hindered C^3=C bond in the higher β-myrcene and β-farnesene monomers and/or from the presence of 2,7-*tert*-butyl substituents on the fluorenyl moiety of catalyst **1**. The presence of a unique T_g for all co- and terpolymers suggests a random distribution of comonomers within sPS segments (consistent with the aforementioned broadening of ^{13}C NMR signals due to increasing presence of styrene-terpene junctions).

As expected, properties of the copolymers were dependent on their composition, which is easily tunable by adjusting the monomer feed ratio. The T_g values significantly decreased with the increase of β-myrcene and β-farnesene contents (Figure 4), and poly(S-*co*-My) materials incorporating between 10 and 20 mol % of β-myrcene units were completely amorphous. As compared with isoprene-based copolymers [17], a wider range of thermal properties is potentially available using these terpenes as comonomers. As shown in Figure 4, the T_g values of the three copolymers were similar at low comonomer incorporation levels (T_g = 70–77 °C for 3–9 mol % of comonomer inserted). For higher amounts of comonomer incorporated (15–30 mol %), the T_g values of poly(S-*co*-My) copolymers monotonously decreased, while those of poly(S-*co*-1,4-*trans*-IP) reached a plateau at ca. 60 °C.

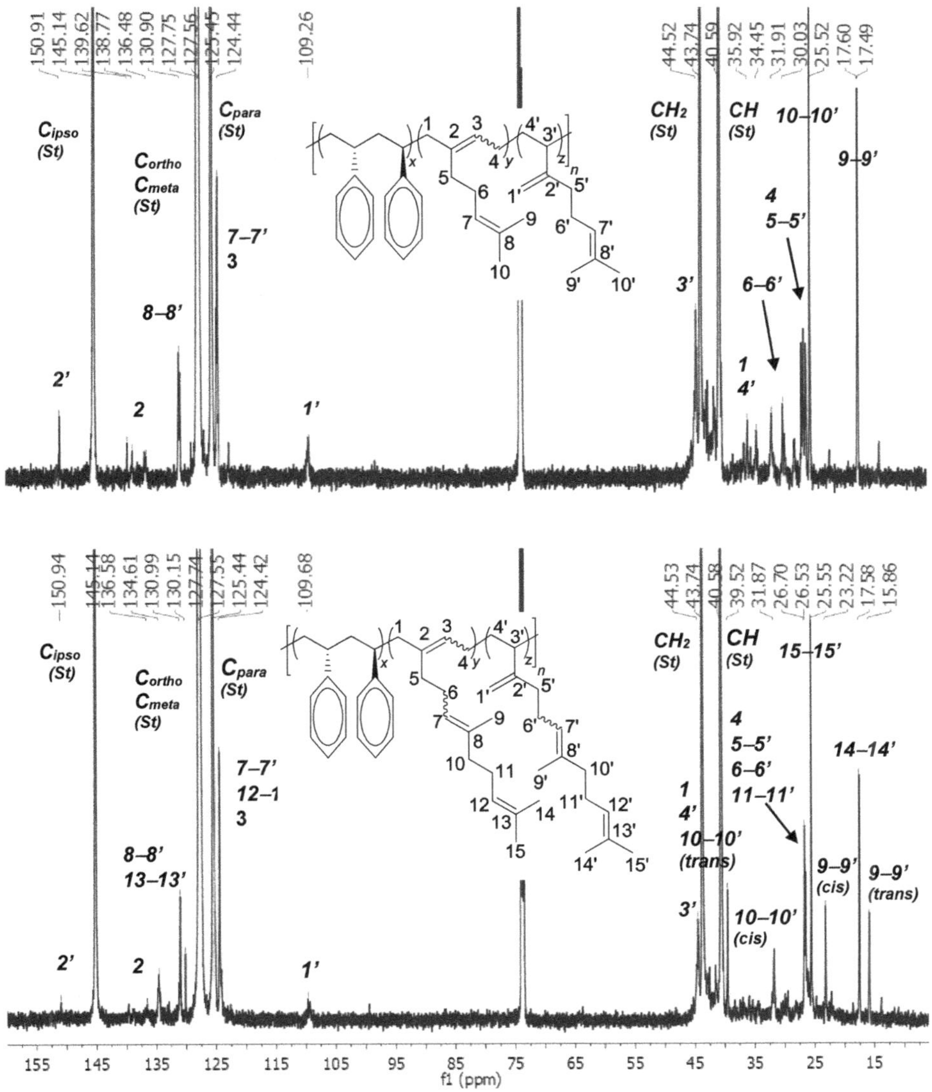

Figure 2. $^{13}C\{^1H\}$ NMR spectra (125 MHz, 1,1,2,2-tetrachloroethane-d_2, 60 °C) of a poly(S-*co*-My) copolymer (top) (Table 1, entry 6), and a poly(S-*co*-Fa) copolymer (bottom) (Table 1, entry 16), with proposed signal assignment.

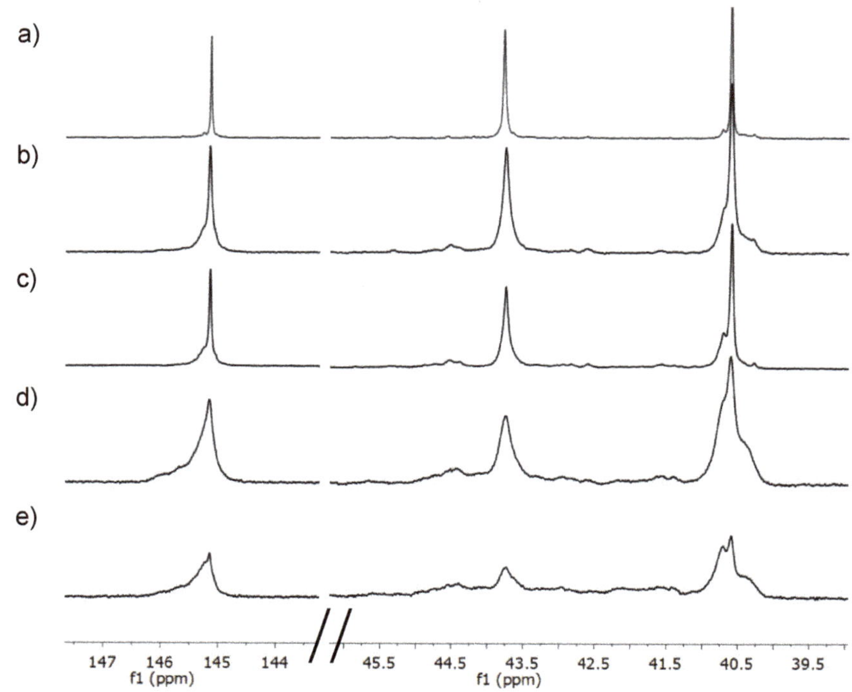

Figure 3. Stack-plot of detailed regions (left, *ipso*; right: methylene carbon) of the $^{13}C\{^1H\}$ NMR spectra (125 MHz, 1,1,2,2-tetrachloroethane-d_2, 60 °C) of poly(S-*co*-My) copolymers (Table 1): (**a**) pure sPS (Table 1, entry 3); (**b**) 5.6 mol % of *β*-myrcene (entry 4); (**c**) 10.9 mol % of *β*-myrcene (entry 6); (**d**) 18.1 mol % of *β*-myrcene (entry 8); (**e**) 30.8 mol % of *β*-myrcene (entry 10).

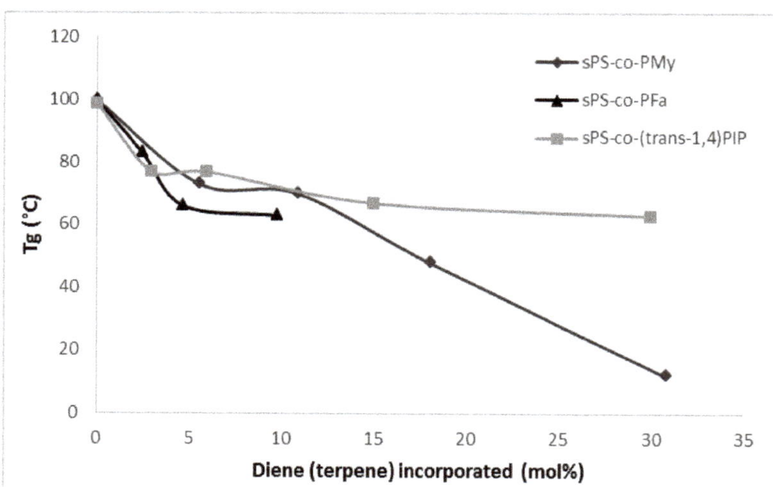

Figure 4. Variation of the glass transition temperature (T_g) of poly(S-*co*-My), poly(S-*co*-Fa) and poly(S-*co*-1,4-*trans*-IP) [17] copolymers as a function of diene (terpene) comonomer content (Table 1, entries 3, 4, 6, 8, 10, 13, 15, 16).

2.3. Terpolymerizations of Styrene with β-Myrcene and Ethylene

A series of styrene/β-myrcene/ethylene terpolymerizations was also performed under the same conditions (Scheme 2, Table 2). The remarkably narrow and monomodal molecular weight distributions suggested the formation of true terpolymers ($Đ_M$ = 1.3–1.6). Complex **1** was moderately productive for styrene/β-myrcene/ethylene terpolymerizations (64–250 kg·mol^{-1}·h^{-1}), and the productivity decreased with an increase in the $[My]_0/[St]_0$ feed ratio. However, it is noteworthy that these productivities were significantly higher than those obtained for styrene/β-myrcene copolymerizations under the same conditions (6–127 kg·mol^{-1}·h^{-1}, compare Tables 1 and 2). This difference in productivity between co- and terpolymerizations is even more pronounced with the increase of $[My]_0/[St]_0$, as productivity values were 2–10 fold larger in terpolymerization than in the corresponding copolymerization. This can be accounted for by the fact that ethylene is a smaller, less sterically demanding monomer than styrene or β-myrcene and, thus, is more easily inserted. Hence, ethylene acts as a "spacer" that facilitates subsequent insertions of styrene or β-myrcene monomer, resulting in higher overall productivities than in copolymerizations without ethylene [22–24]. Under the given conditions, the level of β-myrcene incorporated in terpolymers was equivalent to that observed for styrene/β-myrcene copolymers. In addition, the amount of ethylene incorporated in terpolymers increased (from 18.8 up to 29.2 mol %) with an increase in β-myrcene content.

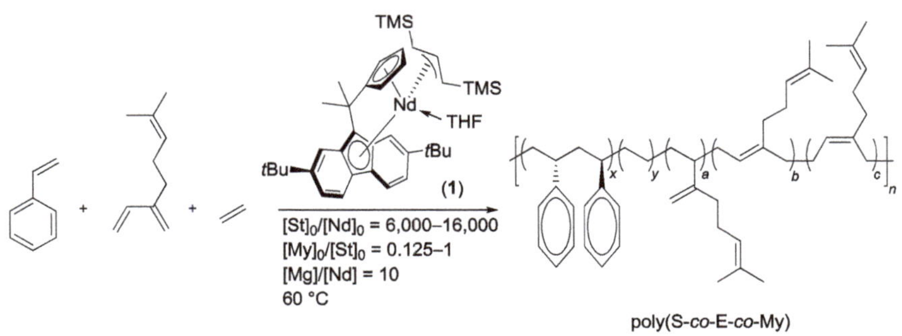

Scheme 2. Styrene/β-myrcene/ethylene terpolymerizations catalyzed by **1**.

Table 2. Styrene/β-myrcene/ethylene terpolymerizations catalyzed by **1** [a].

Entry	$[St]_0$ [M]	$[St]_0/[Ln]$	$[My]_0/[St]_0$	Time [min]	Prod. [b] [kg·mol^{-1}·h^{-1}]	C2 Inc. [c] [mol %]	My Inc. [c] [mol %]	T_m [d] [°C]	T_c [d] [°C]	T_g [d] [°C]	ΔH_m [d] [J·g^{-1}]	$M_n \times 10^{-3}$ [g·mol^{-1}] [e]	$Đ_M$ [e]
1	4.4	8000	0.125	120	250	18.8	4.9	*no*	*no*	61	*no*	38.2	1.31
2 [f]	4.4	16,000	0.125	120	255	19.4	4.3	*no*	*no*	63	*no*	45.3	1.6
3	4.4	8000	0.25	120	167	22.4	9.1	*no*	*no*	58	*no*	35.9	1.29
4	4.4	8000	0.5	180	123	24.2	20.2	*no*	*no*	2	*no*	122.1	1.5
5	3.3	6000	1	360	64	29.2	31.1	*no*	*no*	−21	*no*	87.6	1.5
6 [h]	4.4	8000	-	120	>400	9.5	-	218	151	87	15.0	44.6 [g]	1.9 [g]

[a] General conditions unless otherwise stated: 27–54 μmol of **1**; [Nd] = 5.4 × 10^{-4} mol·L^{-1}; $[Mg(nBu)_2]/[Nd]$ = 10; T_{polym} = 60 °C; $P_{ethylene}$ = 2 bar; styrene and β-myrcene purified through neutral alumina, stirring on CaH_2, trap-to-trap vacuum distillation and stored at −27 °C on 3 Å molecular sieves; *no*: not observed; [b] Productivity calculated over the whole reaction time; [c] Determined by ^{1}H NMR spectroscopy; [d] Determined by DSC from second run; [e] Determined by GPC at 30 °C in THF; [f] [Nd] = 2.7 × 10^{-4} mol·L^{-1}; [g] Determined by GPC at 135 °C in 1,2,4-trichlorobenzene; [h] Styrene-ethylene copolymerization performed under similar conditions.

The poly(S-*co*-E-*co*-My) terpolymers were also characterized by $^{13}C\{^1H\}$ NMR spectroscopy (Supporting Information, Figure S18). As previously observed for styrene/β-myrcene copolymers, PS sequences in the terpolymers were predominantly syndiotactic (see the *ipso* carbon signal at δ 145.2 ppm) and the myrcene units were present as both 1,4- and 3,4-units (for instance, see the resonances at δ 150.9 and 138.6–136.7 ppm assigned to the C2 atom of the backbone of 3,4- and 1,4-insertions, respectively). Again, the T_g values were significantly lower in poly(S-*co*-E-*co*-My) terpolymers than in poly(S-*co*-E-*co*-1,4-*trans*-IP) terpolymers (at a given ethylene incorporation level, ca. 20 mol %) [17]. The T_g values were also lower in poly(S-*co*-E-*co*-My) terpolymers than in poly(S-*co*-My) copolymers containing similar β-myrcene incorporation contents.

3. Materials and Methods

General considerations. All experiments were performed under a dry argon atmosphere, using a glovebox or standard Schlenk techniques. Cyclohexane and *n*-dodecane were distilled from CaH_2 and stored over 3 Å molecular sieves. Toluene was distilled from Na/K alloy and stored over 3 Å molecular sieves. 1,1,2,2-Tetrachloroethane-d_2 (TCE-d_2, 99% D, Acros Organics, Geel, Belgium) was used as received. Styrene (Fisher Chemical, Waltham, MA, USA general purpose grade, stabilized with 10–15 ppm of *tert*-butylcatechol) was eluted through neutral aluminum oxide, stirred and heated over CaH_2, vacuum-distilled and stored over 3 Å molecular sieves at −30 °C under argon. β-Myrcene (Acros Organics, Geel, Belgium, technical grade, stabilized with 0.01% of α-Tocopherol) and farnesene (75:25 α/β mixture of isomers, generously provided by Total Raffinage-Chimie) was dried over CaH_2, transferred by trap-to-trap under vacuum and stored at −30 °C under argon. $(n\mathrm{Bu})_2\mathrm{Mg}$ (1.0 M solution in hexanes, Sigma-Aldrich, Lyon, France) was used as received. Ethylene (Air Liquide, Paris, France, N35) was used without further purification.

Instruments and measurements. ^{1}H and ^{13}C{^{1}H} NMR spectra of co- and terpolymers were recorded on a Bruker AM-500 spectrometer (1,1,2,2-tetrachloroethane-d_2, 60 °C, Bruker, Wissembourg, France). GPC analyses of β-myrcene-based polymers were performed in THF at 30 °C using PS standards for calibration. GPC analyses of farnesene-based polymers were performed in 1,2,4-trichlorobenzene at 135 °C using PS standards for calibration. Differential scanning calorimetry analyses were performed on a Setaram DSC 131 apparatus (SETARAM Instrumentation, Cranbury, NJ, USA), under continuous flow of helium and using aluminum capsules. Crystallization temperatures were measured during the first cooling (10 °C.min^{-1}), and glass transition and melting temperatures were measured during the second heating (10 °C.min^{-1}).

Typical procedure for styrene-terpene copolymerization. In a typical experiment (Table 1, entry 4), complex **1** (ca. 8 mg), cyclohexane (7.6 mL) and $(n\mathrm{Bu})_2\mathrm{Mg}$ (0.1 mL of a 1.0 M solution in hexanes) were introduced in the glovebox in a Schlenk flask. The tube was capped with a septum. Out of the glovebox, the tube was heated with an oil bath at 60 °C. Under vigorous stirring, styrene (9.4 mL) and β-myrcene (1.8 mL) were added with syringes. When the desired polymerization time was reached, methanol (10 mL) was added to quench the reaction. The precipitated polymer was washed with methanol (ca. 50 mL), filtered and dried under vacuum at 60 °C until constant weight was reached.

Typical procedure for styrene/β-myrcene/ethylene terpolymerization. In a typical experiment (Table 2, entry 1), a 300 mL-glass high-pressure reactor (TOP-Industrie, Vaux le Pénil, France) was charged with cyclohexane (41 mL) under argon and heated at the appropriate temperature by circulating water in a double mantle. Under an ethylene flow were introduced styrene (50 mL), β-myrcene (9.4 mL), and $(n\mathrm{Bu})_2\mathrm{Mg}$ (0.5 mL of a 1.0 M solution in hexanes) and the solution of complex **1** in toluene (ca. 43 mg in 2 mL). The ethylene pressure in the reactor was kept constant at 2 bar with a back regulator and the reaction media was mechanically stirred. At the end of the polymerization, the reactor was vented to air, and the copolymer was precipitated in methanol (ca. 500 mL), washed with methanol and dried under vacuum at 60 °C until constant weight was reached.

Calculation of β-myrcene and styrene fractions in poly(S-*co*-My) copolymers. The fraction of β-myrcene and styrene F_{My} and F_{St} in the copolymers was calculated with the following equations:

$$F_{\mathrm{St}} = A_{arom}/5 \tag{1}$$

$$F_{\mathrm{My}} = (A_{aliph} - 3(A_{arom}/5))/13.5 \tag{2}$$

where A_{aliph} is the area of aliphatic hydrogens (δ 0.5–2.5 ppm) and A_{arom} is the area of aromatic hydrogens of styrene (δ 6.5–7.5 ppm). The fraction of β-myrcene can also be determined by considering the area of olefinic hydrogens A_{olefin} (δ 4.0–5.5 ppm) but the uncertainty in the integration of the latter signals is more important than using the aliphatic hydrogens signals.

Calculation of β-farnesene and styrene fractions in poly(S-*co*-Fa) copolymers. The fraction of β-farnesene and styrene F_{Fa} and F_{St} in copolymers was calculated with the following equations:

$$F_{St} = A_{arom}/5 \tag{3}$$

$$F_{Fa} = (A_{aliph} - 3(A_{arom}/5))/20.5 \tag{4}$$

where A_{aliph} is the area of aliphatic hydrogens (δ 0.5–2.5 ppm) and A_{arom} is the area of aromatic hydrogens of styrene (δ 6.5–7.5 ppm).

Calculation of β-myrcene and styrene fractions in poly(S-*co*-E-*co*-My) terpolymers. Styrene, β-myrcene and ethylene fractions in terpolymers were calculated using the following equations:

$$F_{St} = A_{arom}/5 \tag{5}$$

$$F_{My} = A_{olefin}/2.5 \tag{6}$$

$$F_{Eth} = (A_{aliph} - 3(A_{arom}/5) - 13.5(A_{olefin}/2.5))/4 \tag{7}$$

where A_{aliph} is the area of aliphatic hydrogens (δ 0.5–2.5 ppm) and A_{arom} is the area of aromatic hydrogens of styrene (δ 6.5–7.5 ppm).

4. Conclusions

In this contribution, we reported on the production of new poly(S-*co*-My) copolymers and poly(S-*co*-E-*co*-My) terpolymers using a highly productive and syndioselective *ansa*-lanthanidocene catalyst previously described by our group. The incorporation level and the overall catalytic productivity were somewhat lower as compared to analogous styrene/isoprene copolymerizations. The composition of the final polymers was easily controlled by adjusting the initial feed ratio; thus, a wide range of materials with various thermal properties was obtained. In particular, terpene-based co- and terpolymers featured a wider range of T_g values compared with isoprene-based co(ter)polymers already reported by our group (at a given diene incorporation level) [17], indicating the potentially easier processability of those new materials.

Supplementary Materials: Supporting Information is available online at http://www.mdpi.com/2073-4344/7/12/361/s1. Figure S1: ^{1}H NMR spectrum (400 MHz, 25 °C, TCE-d_2) of poly(My) (T_{polym} = 120 °C, Table 1, entry 2), Figure S2: ^{13}C{^{1}H} NMR spectrum (125 MHz, 60 °C, TCE-d_2) of poly(My) (T_{polym} = 60 °C, Table 1, entry 1), Figure S3: ^{13}C{^{1}H} NMR spectrum (100 MHz, 25 °C, TCE-d_2) of poly(My) (T_{polym} = 120 °C, Table 1, entry 2), Figure S4: DSC thermogram of poly(My) (Table 1, entry 1), Figure S5: GPC trace of poly(Fa) (Table 1, entry 1), Figure S6: ^{1}H NMR spectrum (500 MHz, 60 °C, TCE-d_2) of poly(Fa) (Table 1, entry 11), Figure S7: ^{13}C{^{1}H} NMR spectrum (125 MHz, 60 °C, TCE-d_2) of poly(Fa) (Table 1, entry 11), Figure S8: Fineman-Ross plot for the copolymerizations of styrene and myrcene with **1**/(nBu)$_2$Mg at 60 °C and least-square best-fit line (F = [St]/[My] in feed, f = [St]/[My] in copolymer), Figure S9: Typical ^{1}H NMR spectrum (500 MHz, 60 °C, TCE-d_2) of poly(S-*co*-My) (Table 1, entry 8), Figure S10: DSC thermogram of poly(S-*co*-My) (Table 1, entry 8), Figure S11: GPC trace of poly(S-*co*-My) (Table 1, Entry 8), Figure S12: ^{1}H NMR spectra (400 MHz, 25 °C, $CDCl_3$) of farnesene: before styrene-farnesene copolymerization (top), after styrene-farnesene copolymerization (bottom) (Table 1, entry 4), Figure S13: Typical ^{1}H NMR spectrum (500 MHz, 60 °C, TCE-d_2) of poly(S-*co*-Fa) (Table 1, entry 17), Figure S14: Stack-plot of detailed regions (left, ipso; right: methylene carbon) of the ^{13}C{^{1}H} NMR spectra (125 MHz, TCE-d_2, 60 °C) of poly(S-*co*-Fa) copolymers (Table 1): (a) pure sPS (Table 1, entry 3); (b) 2.5 mol % of farnesene (entry 13); (c) 3.4 mol % of farnesene (entry 14); (d) 4.7 mol % of farnesene (entry 16); (e) 9.8 mol % of farnesene (entry 17); Figure S15: DSC thermogram of poly(S-*co*-Fa) (Table 1, entry 14); Figure S16: GPC trace of poly(S-*co*-Fa) (Table 1, entry 14); Figure S17: Typical ^{1}H NMR spectrum (500 MHz, 60 °C, TCE-d_2) of poly(S-*co*-E-*co*-My) (Table 2, entry 3), Figure S18: Typical ^{13}C{^{1}H} NMR spectrum (125 MHz, 60 °C, TCE-d_2) of poly(S-*co*-E-*co*-My) (Table 2, entry 3); Figure S19: DSC thermogram of poly(S-*co*-E-*co*-My) (Table 2, entry 3);

Figure S20. GPC trace of poly(S-*co*-E-*co*-My) (Table 2, entry 3); Figure S21: ^{1}H NMR spectrum (400 MHz, $CDCl_3$, 25 °C) of farnesene as a mixture of α- and β-isomers.

Acknowledgments: This work was gratefully supported by Total SA and Total Research and Technology Feluy (Ph.D. grant to E. Laur).

Author Contributions: Jean-François Carpentier and Evgueni Kirillov designed the study and the experiments. Eva Laur performed the experiments and analyses. Eva Laur, Jean-François Carpentier and Evgueni Kirillov interpreted the experiments and wrote the manuscript. Alexandre Welle, Aurélien Vantomme and Jean-Michel Brusson helped in co-supervision of the study.

Conflicts of Interest: The authors declare no conflict of interest.

References and Notes

1. Behr, A.; Johnen, L. Myrcene as a natural base chemical in sustainable chemistry: A critical review. *ChemSusChem* **2009**, *2*, 1072–1095. [PubMed]
2. Loughmari, S.; Hafid, A.; Bouazza, A.; El Bouadili, A.; Zinck, P.; Visseaux, M. Highly stereoselective coordination polymerization of β-myrcene from a lanthanide-based catalyst: Access to bio-sourced elastomers. *J. Polym. Sci. Part A Polym. Chem.* **2012**, *50*, 2898–2905.
3. Díaz de León Gómez, R.E.; Enríquez-Medrano, F.J.; Maldonado Textle, H.; Mendoza Carrizales, R.; Reyes Acosta, K.; López González, H.R.; Olivares Romero, J.L.; Lugo Uribe, L.E. Synthesis and characterization of high *cis*-polymyrcene using neodymium-based catalysts. *Can. J. Chem. Eng.* **2016**, *94*, 823–832.
4. Liu, B.; Li, L.; Sun, G.; Liu, D.; Li, S.; Cui, D. Isoselective 3,4-(co)polymerization of bio-renewable myrcene using NSN-ligated rare-earth metal precursor: An approach to a new elastomer. *Chem. Commun.* **2015**, *51*, 1039–1041. [CrossRef] [PubMed]
5. Liu, B.; Han, B.; Zhang, C.; Li, S.; Sun, G.; Cui, D. Renewable β-myrcene polymerization initiated by lutetium alkyl complexes ligated by imidophosphonamido ligand. *Chin. J. Polym. Sci.* **2015**, *33*, 792–796.
6. Raynaud, J.; Wu, J.Y.; Ritter, T. Iron-catalyzed polymerization of isoprene and other 1,3-dienes. *Angew. Chem. Int. Ed.* **2012**, *51*, 11805–11808. [CrossRef] [PubMed]
7. Schellenberg, J. *Syndiotactic Polystyrene: Synthesis, Characterization, Processing, and Applications*; John Wiley & Sons: Hoboken, NJ, USA, 2010.
8. Malanga, M. Syndiotactic polystyrene materials. *Adv. Mater.* **2000**, *12*, 1869–1872.
9. Zinck, P.; Bonnet, F.; Mortreux, A.; Visseaux, M. Functionalization of syndiotactic polystyrene. *Prog. Polym. Sci.* **2009**, *34*, 369–392.
10. Jaymand, M. Recent progress in the chemical modification of syndiotactic polystyrene. *Polym. Chem.* **2014**, *5*, 2663–2690.
11. Laur, E.; Kirillov, E.; Carpentier, J.-F. Engineering of syndiotactic and isotactic polystyrene-based copolymers *via* stereoselective catalytic polymerization. *Molecules* **2017**, *22*, 594.
12. Quirk, R.P.; Huang, T.-L. Alkyllithium-initiated polymerization of myrcene: New block copolymers of styrene and myrcene. In *New Monomers and Polymers*; Polymer Science and Technology; Culbertson, B.M., Pittman, C.U., Jr., Eds.; Springer: Boston, MA, USA, 1984; Volume 25, pp. 329–355.
13. Sarkar, P.; Bhowmick, A.K. Terpene based sustainable elastomer for low rolling resistance and improved wet grip application: Synthesis, characterization and properties of poly(styrene-*co*-myrcene). *ACS Sustain. Chem. Eng.* **2016**, *4*, 5462–5474. [CrossRef]
14. Métafiot, A.; Kanawati, Y.; Gérard, J.-F.; Defoort, B.; Marić, M. Synthesis of β-myrcene-based polymers and styrene block and statistical copolymers by SG1 nitroxide-mediated controlled radical polymerization. *Macromolecules* **2017**, *50*, 3101–3120. [CrossRef]
15. Georges, S.; Touré, A.O.; Visseaux, M.; Zinck, P. Coordinative chain transfer copolymerization and terpolymerization of conjugated dienes. *Macromolecules* **2014**, *47*, 4538–4547. [CrossRef]
16. Laur, E.; Louyriac, E.; Dorcet, V.; Welle, A.; Vantomme, A.; Miserque, O.; Brusson, J.-M.; Maron, L.; Carpentier, J.-F.; Kirillov, E. Substitution effects in highly syndioselective styrene polymerization catalysts based on single-component allyl *ansa*-lanthanidocenes: An experimental and theoretical study. *Macromolecules* **2017**, *50*, 6539–6551.

17. Rodrigues, A.-S.; Kirillov, E.; Vuillemin, B.; Razavi, A.; Carpentier, J.-F. Stereocontrolled Styrene–isoprene copolymerization and styrene–ethylene–isoprene terpolymerization with a single-component allyl *ansa*-neodymocene catalyst. *Polymer* **2008**, *49*, 2039–2045. [CrossRef]
18. Georges, S.; Bria, M.; Zinck, P.; Visseaux, M. Polymyrcene microstructure revisited from precise high-field nuclear magnetic resonance analysis. *Polymer* **2014**, *55*, 3869–3878. [CrossRef]
19. Newmark, R.A.; Majumdar, R.N. ^{13}C-NMR spectra of *cis*-polymyrcene and *cis*-polyfarnesene. *J. Polym. Sci. Part A Polym. Chem.* **1988**, *26*, 71–77.
20. Fineman, M.; Ross, S.D. Linear method for determining monomer reactivity ratios in copolymerization. *J. Polym. Sci.* **1950**, *5*, 259–262. [CrossRef]
21. Due to the fact that the ratio of α/β isomers of farnesene in the monomer was not exactly known and that only the β-isomer was polymerized, the reactivity ratios related to the styrene-farnesene copolymerization could not be determined.
22. Tan, R.; Guo, F.; Li, Y. Copolymerization of propylene with styrene and ethylene by a THF-containing half-sandwich scandium catalyst: Efficient synthesis of polyolefins with a controllable styrene content. *Polym. Chem.* **2017**, *4*, 482–489. [CrossRef]
23. Li, X.; Hou, Z. Scandium-catalyzed copolymerization of ethylene with dicyclopentadiene and terpolymerization of ethylene, dicyclopentadiene, and styrene. *Macromolecules* **2005**, *38*, 6767–6769.
24. Guo, F.; Nishiura, M.; Koshino, H.; Hou, Z. Cycloterpolymerization of 1,6-heptadiene with ethylene and styrene catalyzed by a THF-free half-sandwich scandium complex. *Macromolecules* **2011**, *44*, 2400–2403. [CrossRef]

catalysts

MDPI

Article

Synthesis of Stereodiblock Polybutadiene Using $Cp^*Nd(BH_4)_2(thf)_2$ as a Catalyst

Ryo Tanaka *, Yuto Shinto, Yuushou Nakayama and Takeshi Shiono

Graduate School of Engineering, Hiroshima University, 1-4-1 Kagamiyama, Higashi-hiroshima 739-8527, Japan; m161905@hiroshima-u.ac.jp (Y.S.); yuushou@hiroshima-u.ac.jp (Y.N.); tshiono@hiroshima-u.ac.jp (T.S.)

* Correspondence: rytanaka@hiroshima-u.ac.jp; Tel.: +81-82-424-7729

Received: 7 September 2017; Accepted: 19 September 2017; Published: 25 September 2017

Abstract: Butadiene polymerization, in both a highly *cis*- and *trans*-specific manner, was achieved by using a $Cp^*Nd(BH_4)_2(thf)_2$–Bu_2Mg system as an initiator. The *cis-/trans-* ratio can be tuned by the amount of trialkylaluminum-depleted modified methylaluminoxane (dMMAO). The *cis*-regularity of the polymer was much higher than those obtained by $Nd(BH_4)_3(thf)_3$. The molecular weight of *cis*-regular polymer was increased according to polymer yield, showing that there was no termination or chain transfer reaction during the polymerization. Synthesis of stereodiblock polybutadiene, which showed a high melting temperature (T_m) compared with stereodiblock polyisoprene, was also performed by the addition dMMAO during the polymerization.

Keywords: coordination polymerization; neodymium catalyst; polyconjugated dienes; stereoblock polymer; rubber

1. Introduction

Polyconjugated dienes change their thermal and mechanical properties according to the stereoregularity. Generally, *cis*-1,4-regular polymer shows typical elastomeric property with a very low glass transition temperature (T_g) and *trans*-1,4-regular polymer has a melting point (T_m). Based on these properties, synthetic *cis*-1,4 polymer is widely applied as an alternative material to natural rubber, especially in tire industries [1]. *Trans*-1,4 polymer has great potential for shape-memory rubber and blend in *cis*-rich polymer which improves mechanical properties such as the rolling resistance of tires. Stereoblock polyconjugated dienes, which possess both *cis*- and *trans*-sequences in a single polymer chain, would be a promising material showing unique thermal and mechanical properties.

Some synthetic examples of stereoblock polymers are reported using Ni [2], Co [3,4] and Fe [5] catalysts with additional ligands, or rare-earth metal catalysts with chain transfer reagent [6]. However, these stereoblock polymers always contain atactic blocks or short stereoregular blocks consist of less than 10 repeating units. Highly stereoregular long blocks are difficult to synthesize, probably because of the difficulty of changing the stereoselectivity during the polymerization whilst keeping the living manner.

Previously, we reported the synthesis of stereodiblock polyisoprene which consists of *cis*-1,4- and *trans*-1,4-sequences using a Nd/Mg/Al combined catalyst system (Scheme 1) [7]. This is based on the idea that a simple neodymium catalyst system such as $Nd(BH_4)_3(thf)_3/Bu_2Mg$ and $(C_3H_5)NdCl_2$/methylaluminoxane (MAO) can promote *trans*- and *cis*-specific polymerization of isoprene and butadiene in a living manner, respectively (Figure 1) [8–14]. The stereospecificity is basically strongly affected by the metal charge; namely, neutral systems give *trans*-polymer and cationic systems give *cis*-polymer, and the change of the metal charge from neutral to cationic during the polymerization with Lewis acid gave stereodiblock polymer. The strategy would be applied to the other conjugated diene monomers, and as an extensive example, we attempted to synthesize stereoblock polybutadiene herein. The T_m of the polymer from the *trans*-1,4 polybutadiene block, which is an important property for the application to the thermoplastic elastomer, was significantly higher than that of stereodiblock polyisoprene.

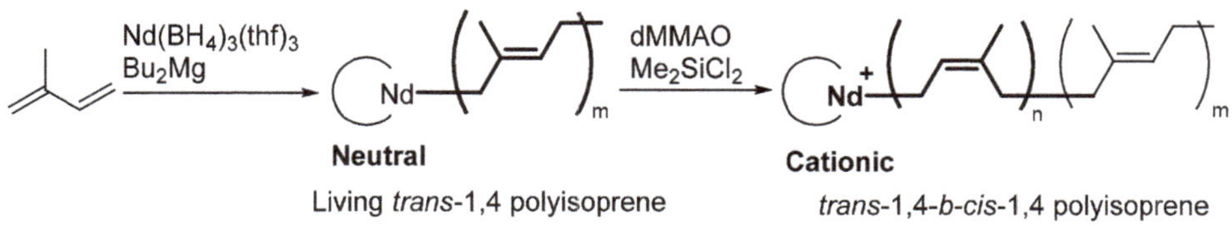

Scheme 1. Synthesis of stereoblock polyisoprene using $Nd(BH_4)_3(thf)_3$ as a catalyst precursor. dMMAO: trialkylaluminum-depleted modified methylaluminoxane.

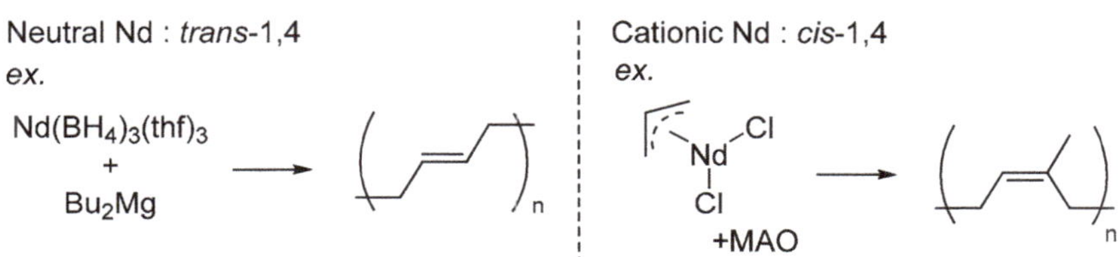

Figure 1. Examples of living butadiene/isoprene polymerization using simple neodymium catalysts. MAO: methylaluminoxane.

2. Results and Discussion

Butadiene polymerization using a combination of $Cp^*Nd(BH_4)_2(thf)_2$ and Bu_2Mg was performed in the presence of excess dMMAO (Table 1). In the previous research, $^iBu_3Al$ was an efficient chain transfer agent of borohydrido-neodymium catalyzed polymerization [15,16], and we therefore removed free trialkylaluminums from MMAO to prevent chain transfer reaction, which is a critical problem for block polymer synthesis. It is already reported that $Cp^*Nd(BH_4)_2(thf)_2$–Bu_2Mg promotes *trans*-1,4-specific polymerization of butadiene with narrow molecular weight distribution, and we successfully reproduced it (Run 1). The addition of 25 equivalents of dMMAO accelerated the reaction so that an almost quantitative amount of polymer was obtained at room temperature within an hour (Run 2). To reproduce the polymerization well, pre-activation of $Cp^*Nd(BH_4)_2(thf)_2$ with Bu_2Mg in the presence of a small amount of butadiene was required. Precipitation of the catalyst occurred without the addition of monomer, which indicated that formation of allylneodymium species was important. The molecular weight distribution was larger than the polymer obtained without dMMAO, but it is probably because of the high viscosity of the reaction mixture. The *trans*-specificity was greatly reduced according to the amount of dMMAO and finally *cis*-specificity reached 93% (Runs 2, 4, and 5), although the activity was greatly reduced with a high Al/Nd ratio. The increase of the Bu_2Mg amount slightly lowered *cis*-specificity (Runs 3 and 4). These tendencies of stereospecificity were similar with those observed in isoprene polymerization using the Nd–Mg/Al system, and use of $Cp^*Nd(BH_4)_2(thf)_2$ showed much higher *cis*-specificity and narrower molecular weight distribution than the $Nd(BH_4)_3(thf)_3$–Bu_2Mg system (Runs 4 and 6).

The ^{13}C NMR spectrum of the polymer showed four distinct signals around 130 ppm, which is assigned to stereodyad sequences (Figure 2, see also the Supplementary Materials) [17]. From the integral ratios of these four signals, the ratio of stereodyad (*tt*:*tc*:*ct*:*cc* = 17:21:22:40) was close to the calculated statistical distribution (*tt*:*tc*:*ct*:*cc* = 14:24:24:38) supposing that *trans*/*cis* selectivity was 38% and 62%, which was the same as the polymerization result, respectively. This result showed randomly distributed *trans* and *cis* sequences in the obtained polymer and the occurrence of interconversion between *cis*-specific and *trans*-specific active species via disproportionation (Figure 3). The increase of *cis*-specificity along with the Al/Mg ratio showed that the neutral active species with *trans*-specificity was gradually converted to a cationic *cis*-specific one by the increasing amount of dMMAO, similar to $Nd(BH_4)_3(thf)_3$-catalyzed isoprene polymerization previously reported.

Table 1. Butadiene polymerization using the $Cp^*Nd(BH_4)_2(thf)_2$–Mg/Al catalyst system.

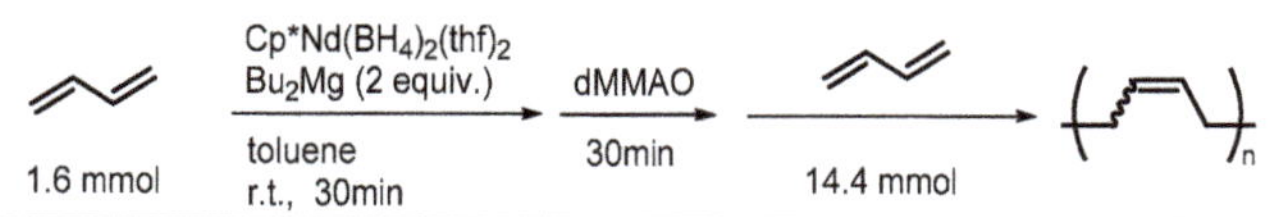

Run	Nd (μmol)	Al/Mg (mol/mol)	Time (h)	Yield (%)	M_n [a] ($\times 10^4$)	M_w/M_n [a]	*trans*:*cis*:vinyl [b] (mol %)
1 [c]	25	0	1	20	0.8	1.3	93:3:4
2	25	25	1	85	8.2	1.5	34:61:5
3 [d]	25	25	0.5	70	7.9	1.7	15:81:4
4	25	50	1	84	10.2	1.7	13:84:3
5	25	100	12	28	2.7	1.7	6:92:2
6 [e]	50	25	1.5	93	5.2	2.5	59:38:3

[a] Determined by GPC (gel permiation chromatography) calibrated with PS standard. [b] Determined by ^1H and ^{13}C NMR. [c] Performed at 40 °C. [d] 4 equivalents of Bu_2Mg was used. [e] $Nd(BH_4)_3(thf)_3$ was used instead of $Cp^*Nd(BH_4)_2(thf)_2$.

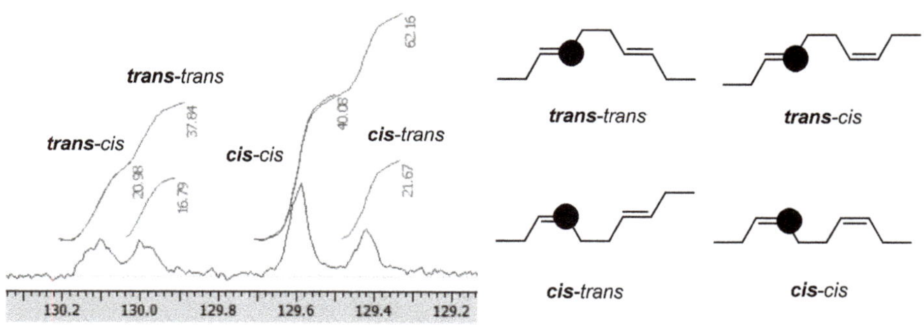

Figure 2. ^{13}C NMR spectrum of polybutadiene (Table 1, Run 2, 125 MHz, in $CDCl_3$).

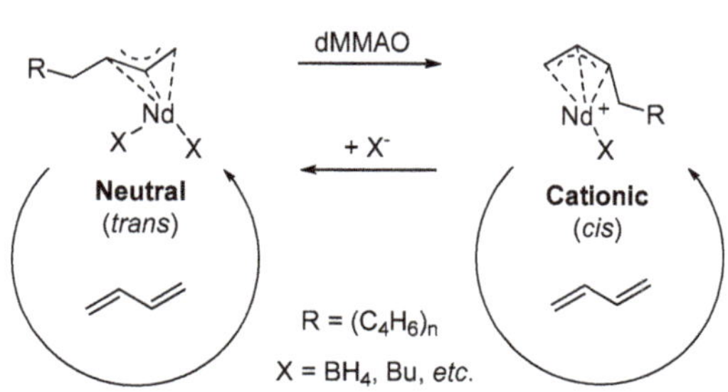

Figure 3. Plausible mechanism of stereospecificity change of butadiene polymerization at low dMMAO concentration.

Next, the relationship between the initial butadiene feed and molecular weight of the polymer was investigated (Table 2). Each run did not reach full conversion because of the high viscosity of the reaction mixture at the end of the polymerization and gave relatively broad molecular weight distribution. However, a constant *N* value, which is the number of polymer chains per Nd catalyst, showed that the molecular weight of the polybutadiene was linearly increased according to the amount of consumed monomer. These results indicated that no chain transfer reaction exists during the polymerization, which is a critical problem for synthesizing the block polymer.

Table 2. Effect of initial monomer feed on the molecular weight in the butadiene polymerization using the $Cp^*Nd(BH_4)_2(thf)_2$–Mg/Al catalyst system.

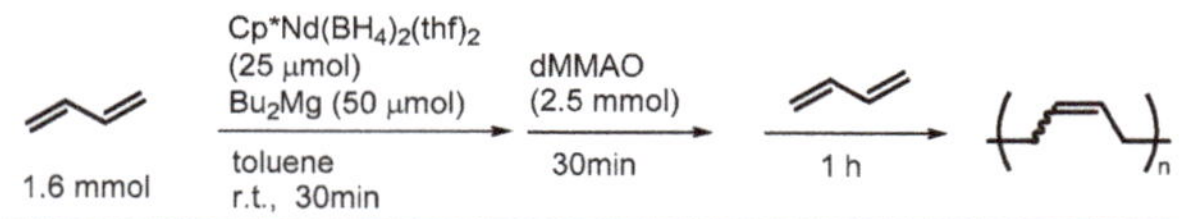

Run	Bd Total (mmol)	Bd/Nd (mol/mol)	Yield (%)	M_n [a] ($\times 10^4$)	PDI [a]	*Trans*:*Cis*:Vinyl [b] (mol %)	N [c]
7	8	320	98	4.9	1.9	14:83:3	0.35
8	16	640	84	10.2	1.7	13:84:3	0.28
9	32	1280	73	16.0	1.7	20:76:4	0.32

[a] Determined by GPC; calibrated with PS standard. [b] Determined by ^1H and ^{13}C NMR. [c] Number of polymer chains per Nd calculated from M_n and yield. Bd: butadiene.

Synthesis of stereodiblock polybutadiene was performed by the addition of dMMAO during *trans*-specific polymerization using the $Cp^*Nd(BH_4)_2(thf)_2$–Bu_2Mg system (Scheme 2). The *cis/trans* ratio was not proportional to the molecular weight, probably because the generation of *cis*-specific active species is slow, but the stereodiblock polymer with a narrow molecular weight distribution was obtained. The whole GPC trace of the *trans*-prepolymer was shifted to a higher molecular weight region in the stereodiblock polymer, indicating high block efficiency (Figure 4). The *trans–trans* and *cis–cis* stereodyad ratio calculated from the ^{13}C NMR spectrum was much higher than the ratio of the others (Figure 5). Moreover, the polymer was not soluble in Et_2O, showing that there was no *cis*-polybutadiene included in the obtained polymer. The difference of T_m and T_g of the polymer measured by DSC (T_m = 50 °C, T_g = −103 °C, Figure 6) was broader than that of stereodiblock polyisoprene [7] (T_m = 32 °C, T_g = −67 °C), showing potential as a building block for the thermoplastic elastomer.

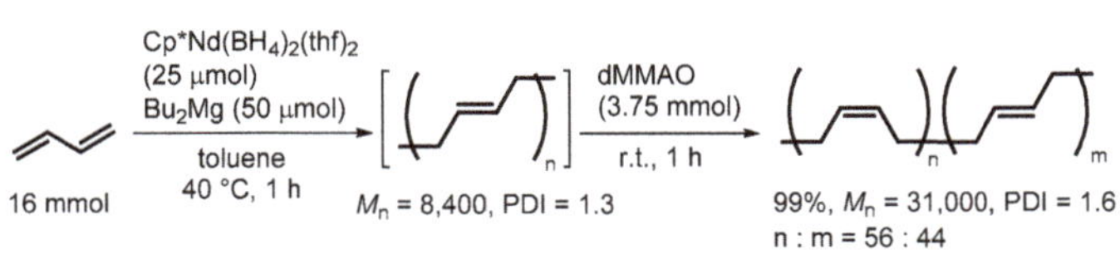

Scheme 2. Synthesis of stereoblock polybutadiene using $Cp^*Nd(BH_4)_2(thf)_2$ (Run 10).

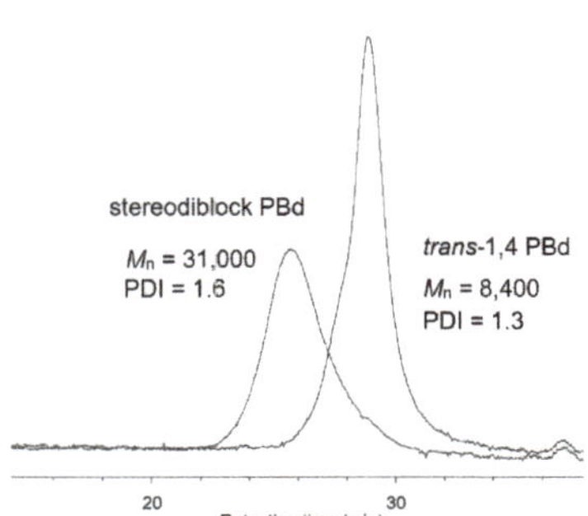

Figure 4. GPC traces of *trans*-1,4 prepolymer and stereodiblock polymer synthesized in Run 10.

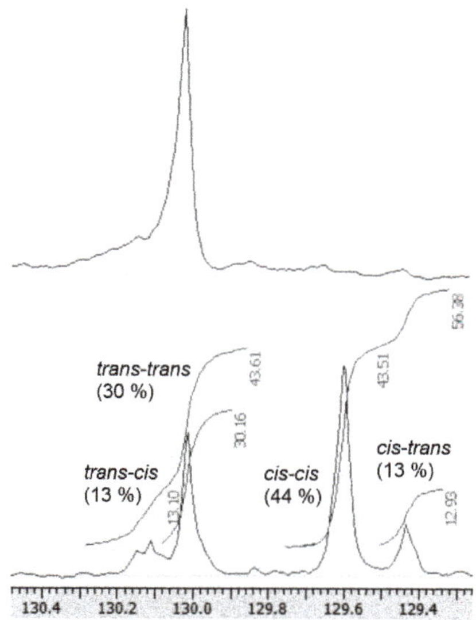

Figure 5. ^{13}C NMR spectrum (125 MHz, in $CDCl_3$) of *trans*-1,4 prepolymer (**above**) and stereodiblock polybutadiene synthesized in Scheme 2 (**below**).

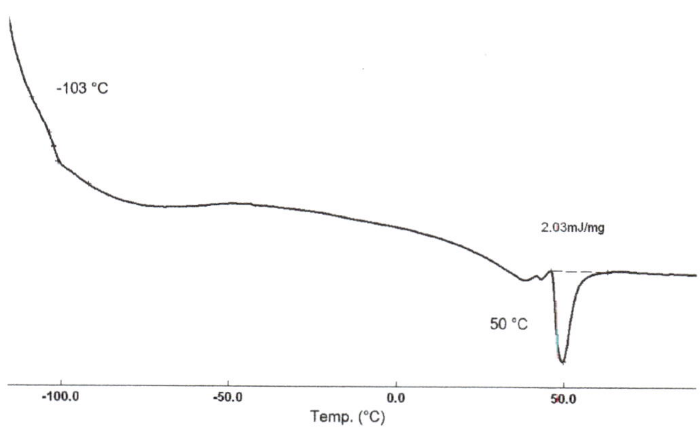

Figure 6. Differential scanning calorimetry (DSC) trace of stereoblock polybutadiene synthesized in Run 10.

3. Experimental Section

3.1. General

All manipulations were performed under an atmosphere of nitrogen using standard Schlenk line techniques. Trialkylaluminum-depleted MMAO (dMMAO) was prepared by the treatment of modified methylaluminoxane (MMAO), which was generously donated by Tosoh-Finechem Co. (Shunan, Japan), with SiO_2 according to the literature [18]. Dry toluene was purchased from Kanto Chemical Co. Inc. (Tokyo, Japan), and a trace of residual water was removed by reaction with sodium metal. Butadiene solution in toluene were purchased from Tokyo Chemical Industry Co. Ltd. (Tokyo, Japan) and used as received. $Nd(BH_4)_3(thf)_3$ and $Cp^*Nd(BH_4)_2(thf)_2$ was prepared according to the literature and used immediately after the preparation [19]. Other materials were used without further purifications.

^1H and ^{13}C nuclear magnetic resonance (NMR) spectra were recorded in chloroform-*d* on a Varian 500 NMR spectrometer. The obtained spectra were referenced to the signal of residual protonated solvent [^1H: δ = 7.26 ppm] or the signal of solvent [^{13}C: δ = 77.16 ppm]. Molecular weights of polymers were determined by Tosoh HLC-8320 gel permiation chromatography (GPC) system (Tokyo, Japan), calibrated with polystyrene standard (T = 40 °C; eluent: THF). T_g, T_m and melting enthalpy (ΔH_m) of the polymer were measured by differential scanning calorimetry (DSC) analyses performed by a SII EXSTAR6000 system (Tokyo, Japan).

3.2. Polymerization Procedure

3.2.1. Butadiene Polymerization Using the Cp*Nd(BH_4)$_2$(thf)$_2$–Bu_2Mg/dMMAO System

To a 20 mL Schlenk flask, Cp*Nd(BH_4)$_2$(thf)$_2$ (11 mg, 25 μmol) was charged and dissolved into butadiene solution in toluene (0.50 mL, containing 1.6 mmol butadiene). To this light blue solution, Bu_2Mg in heptane (50 μL, 1.0 M, 50 μmol) was added and the resulting yellow-green solution was stirred at room temperature for 30 min. dMMAO solution in toluene (1.10 mL, 1.25 mmol) was added to the solution and further stirred at room temperature for 30 min. To the resulting light-yellow solution, butadiene in toluene (3.2 M, 4.5 mL, 14.4 mmol) was added to start the polymerization. After stirring at room temperature for 30 min, the reaction mixture was poured into acidic methanol containing 2% of hydrochloric acid and the precipitated solid was recovered. The polymer was dried under vacuum overnight until constant weight. An amount of 735 mg (85%) of colorless viscous polymer was obtained.

3.2.2. Synthesis of Stereoblock Polybutadiene Using Cp*Nd(BH_4)$_2$(thf)$_2$ as a Catalyst

To a 20 mL Schlenk flask, Cp*Nd(BH_4)$_2$(thf)$_2$ (11 mg, 25 μmol), butadiene solution in toluene (3.2 M, 5.0 mL, 16 mmol) and Bu_2Mg in heptane (50 μmol, 1.0 M, 50 μL) was charged and stirred for 1 h at 40 °C. The resulting mixture was cooled to room temperature and a toluene solution of dMMAO (3.75 mL, 3.75 mmol) was added. The mixture was further stirred for 1 h and poured into acidic methanol containing 2% of hydrochloric acid. The polymer was recovered and dried under vacuum overnight until constant weight. An amount of 862 mg (99%) of colorless polymer was obtained.

3.2.3. Determination of Stereoregularity of Polybutadiene

The ratio of 1,4- and 1,2-unit was calculated from ^1H NMR spectra according to the following equation [17]:

$$R_{1,2}(\%) = 200I_{ter}/(2I_{int} + I_{ter})$$

where $R_{1,2}$ represents the ratio of 1,2-unit, I_{int} represents the integral ratio of internal olefinic protons observed at 5.4 ppm, and I_{ter} represents the integral ratio of terminal vinyl protons observed at 5.0 ppm. The ratio of *trans*-1,4 and *cis*-1,4 unit was calculated from the integral ratio of the signals at 27.4 (*cis*-1,4) and 32.7 ppm (*trans*-1,4) on ^{13}C NMR spectra.

4. Conclusions

Stereodiblock polybutadiene was successfully synthesized by using Cp*Nd(BH_4)$_2$(thf)$_2$ as a catalyst. Unlike the previously reported synthesis of stereodiblock polyisoprene using Nd(BH_4)$_3$(thf)$_3$, there was no need to add a chloride source such as ^{t}BuCl and Me_2SiCl_2. The obtained stereodiblock polybutadiene showed high T_m compared with the corresponding stereodiblock polyisoprene. This result showed the great potential of stereoblock polydienes for application to the thermoplastic elastomer.

Supplementary Materials: The following are available online at www.mdpi.com/2073-4344/7/10/284/s1. Figure S1. ^1H NMR spectrum of polybutadiene obtained in Table 1, Run 1 (500 MHz, in $CDCl_3$), Figure S2. ^{13}C NMR spectrum of polybutadiene obtained in Table 1, Run 1 (125 MHz, in $CDCl_3$), Figure S3. ^1H NMR spectrum of polybutadiene obtained in Table 1, Run 2 (500 MHz, in $CDCl_3$), Figure S4. ^{13}C NMR spectrum of polybutadiene obtained in Table 1, Run 2 (125 MHz, in $CDCl_3$), Figure S5. ^{1}H NMR spectrum of polybutadiene obtained in Table 1, Run 3 (500 MHz, in $CDCl_3$), Figure S6. ^{13}C NMR spectrum of polybutadiene obtained in Table 1, Run 3 (125 MHz, in $CDCl_3$), Figure S7. ^{1}H NMR spectrum of polybutadiene obtained in Table 1, Run 4 (500 MHz, in $CDCl_3$), Figure S8. ^{13}C NMR spectrum of polybutadiene obtained in Table 1, Run 4 (125 MHz, in $CDCl_3$), Figure S9. ^{1}H NMR spectrum of polybutadiene obtained in Table 1, Run 5 (500 MHz, in $CDCl_3$), Figure S10. ^{13}C NMR spectrum of polybutadiene obtained in Table 1, Run 5 (125 MHz, in $CDCl_3$), Figure S11. ^{1}H NMR spectrum of polybutadiene obtained in Table 1, Run 6 (500 MHz, in $CDCl_3$), Figure S12. ^{13}C NMR spectrum of polybutadiene obtained in Table 1, Run 6 (125 MHz, in $CDCl_3$), Figure S13. ^{1}H NMR spectrum of polybutadiene obtained in Table 2, Run 7 (500 MHz, in $CDCl_3$), Figure S14. ^{13}C NMR spectrum of polybutadiene obtained in Table 2, Run 7 (125 MHz, in $CDCl_3$), Figure S15. ^{1}H NMR spectrum of polybutadiene obtained in Table 2, Run 9 (500 MHz, in $CDCl_3$), Figure S16. ^{13}C NMR spectrum of polybutadiene obtained in Table 2, Run 3 (125 MHz, in $CDCl_3$), Figure S17. ^{1}H NMR spectrum of stereoblock polybutadiene (500 MHz, in $CDCl_3$), Figure S18. ^{13}C NMR spectrum of stereoblock polybutadiene (125 MHz, in $CDCl_3$).

Acknowledgments: The authors thank Tosoh-Finechem Co. for generous donation of MMAO.

Author Contributions: Ryo Tanaka conceived and designed the experiments; Yuto Shinto performed the experiments and a part of them were reproduced by Ryo Tanaka; Ryo Tanaka, Yuushou Nakayama and Takeshi Shiono analyzed the data; Ryo Tanaka wrote the paper.

Conflicts of Interest: The authors declare no conflict of interest.

References

1. Brandt, H.D.; Nentwig, W.; Rooney, N.; LaFlair, R.T.; Wolf, U.U.; Duffy, J.; Puskas, J.E.; Kaszas, G.; Drewitt, M.; Glander, S. Rubber, 5. Solution Rubbers. In *Ullmann's Encyclopedia of Industrial Chemistry*, 7th ed.; Elvers, B., Ed.; Wiley-VCH: Weinheim, Germany, 2011.
2. Hadjiandreou, P.; Julemont, M.; Teyssie, P. Butadiene 1,4-Polymerization Initiated by Bis[(η^3-allyl)(trifluoroacetato)nickel]: A Perfectly "Living" Coordination System. *Macromolecules* **1984**, *17*, 2455–2456. [CrossRef]
3. Cai, Z.; Shinzawa, M.; Nakayama, Y.; Shiono, T. Synthesis of Regioblock Polybutadiene with $CoCl_2$-Based Catalyst via Reversible Coordination of Lewis Base. *Macromolecules* **2009**, *42*, 7642–7643. [CrossRef]
4. Tanaka, R.; Kasai, Y.; Shinzawa, M.; Cai, Z.; Nakayama, Y.; Shiono, T. Synthesis of multiblock copolymer of poly(*cis*-1,4-butadiene) and poly(3-buten-1-ol). *Macromol. Chem. Phys.* **2014**, *215*, 888–892. [CrossRef]
5. Zheng, W.; Wang, F.; Bi, J.; Zhang, H.; Zhang, C.; Hu, Y.; Bai, C.; Zhang, X. Synthesis and Characterization of Soft–Hard Stereoblock Polybutadiene with Fe(2-EHA)$_3$/Al(*i*-Bu)$_3$/DEP Catalyst System. *J. Polym. Chem. Part A Polym. Chem.* **2015**, *53*, 1182–1188. [CrossRef]
6. Phuphuak, Y.; Bonnet, F.; Stoclet, G.; Briad, M.; Zinck, P. Isoprene chain shuttling polymerisation between *cis* and *trans* regulating catalysts: Straightforward access to a new material. *Chem. Commun.* **2017**, *53*, 5330–5333. [CrossRef] [PubMed]
7. Tanaka, R.; Yuuya, K.; Sato, H.; Eberhardt, P.; Nakayama, Y.; Shiono, T. Synthesis of stereodiblock polyisoprene consisting of *cis*-1,4 and *trans*-1,4 sequences by using a neodymium catalyst: Change of the stereospecificity triggered by an aluminum compound. *Polym. Chem.* **2016**, *7*, 1239–1243. [CrossRef]
8. Zhang, Z.; Cui, D.; Wang, B.; Liu, B.; Yang, Y. Polymerization of 1,3-Conjugated Dienes with Rare-Earth Metal Precursors. *Mol. Catal. Rare-Earth Elem.* **2010**, *137*, 49–108. [CrossRef]
9. Maiwald, S.; Weißenborn, H.; Sommer, C.; Müller, G.; Taube, R. Komplexkatalyse LIX. Die Katalyse der 1,4-*trans*-Polymerisation des Butadiens mit Tris(allyl)neodym(III) Nd(η^3-C_3H_5)$_3$ als Einkomponenten katalysator—Kinetik und Reaktionsmechanismus. *J. Organomet. Chem.* **2001**, *640*, 1–9. [CrossRef]
10. Bonnet, F.; Visseaux, M.; Pereira, A.; Barbier-Baudry, D. Highly *trans*-Stereospecific Isoprene Polymerization by Neodymium Borohydrido Catalysts. *Macromolecules* **2005**, *38*, 3162–3169. [CrossRef]
11. Ventura, A.; Chenal, T.; Bria, M.; Bonnet, F.; Zinck, P.; Ngono-Ravache, Y.; Balanzat, E.; Visseaux, M. *Trans*-stereospecific polymerization of butadiene and random copolymerization with styrene using borohydrido neodymium/magnesium dialkyl catalysts. *Eur. Polym. J.* **2013**, *49*, 4130–4140. [CrossRef]
12. Fadlallah, S.; Terrier, M.; Jones, C.; Roussel, P.; Bonnet, F.; Visseaux, M. Mixed Allyl–Borohydride Lanthanide Complexes: Synthesis of Ln(BH_4)$_2$(C_3H_5)(THF)$_3$ (Ln = Nd, Sm), Characterization, and Reactivity toward Polymerization. *Organometallics* **2016**, *35*, 456–461. [CrossRef]

13. Maiwald, S.; Sommer, C.; Müller, G.; Taube, R. On the 1,4-*cis*-Polymerization of Butadiene with the Highly Active Catalyst Systems $Nd(C_3H_5)_2Cl{\cdot}1.5$ THF/Hexaisobutylaluminoxane (HIBAO), $Nd(C_3H_5)Cl_2{\cdot}2$ THF/HIBAO and $Nd(C_3H_5)Cl_2{\cdot}2$ THF/Methylaluminoxane (MAO)—Degree of Polymerization, Polydispersity, Kinetics and Catalyst Formation. *Macromol. Chem. Phys.* **2001**, *202*, 1446–1456.
14. Ajellal, N.; Furlan, L.; Thomas, C.M.; Casagrande, O.L., Jr.; Carpentier, J.F. Mixed Aluminum-Magnesium-Rare Earth Allyl Catalysts for Controlled Isoprene Polymerization: Modulation of Stereocontrol. *Macromol. Rapid Commun.* **2006**, *27*, 338–343. [CrossRef]
15. Conti, F.; Segre, A.; Pini, P.; Porri, L. Nuclear magnetic resonance studies of polydienes: 1. ^{13}C n.m.r, of 1,4-polybutadiene obtained by π-allyl nickel trifluoroacetate catalysts. *Polymer* **1974**, *15*, 5–8. [CrossRef]
16. Visseaux, M.; Mainil, M.; Terrier, M.; Mortreux, A.; Roussel, P.; Mathivet, T.; Destarac, M. Cationic borohydrido–neodymium complex: Synthesis, characterization and its application as an efficient pre-catalyst for isoprene polymerization. *Dalton Trans.* **2008**, 4558–4561. [CrossRef]
17. Valente, A.; Zinck, P.; Mortreux, A.; Visseaux, M. Borohydrido Rare Earth Based Coordinative Chain Transfer Copolymerization: A Convenient Tool for Tuning the Microstructure of Isoprene/Styrene Copolymers. *J. Polym. Sci. Part A Polym. Chem.* **2011**, *49*, 1615–1620. [CrossRef]
18. Tanaka, R.; Kawahara, T.; Shinto, Y.; Nakayama, Y.; Shiono, T. An alternative method for the preparation of trialkylaluminum-depleted modified-methylaluminoxane (dMMAO). *Macromolecules* **2017**, *50*, 5989–5993. [CrossRef]
19. Cendrowski-Guillaume, S.M.; Le Gland, G.; Nierlich, M.; Ephritikhine, M. Lanthanide Borohydrides as Precursors to Organometallic Compounds. Mono(cyclooctatetraenyl) Neodymium Complexes. *Organometallics* **2000**, *19*, 5654–5660. [CrossRef]

catalysts

MDPI

Review

Recent Advances in Rare Earth Complexes Bearing Allyl Ligands and Their Reactivity towards Conjugated Dienes and Styrene Polymerization

Jashvini Jothieswaran, Sami Fadlallah, Fanny Bonnet * and Marc Visseaux *

UMR 8181—UCCS—Unité de Catalyse et de Chimie du Solide, ENSCL, Centrale Lille, University Artois, University Lille, CNRS, F-59000 Lille, France; s1438669@sms.ed.ac.uk (J.J.); sami.fadlallah@etudiant.univ-lille1.fr (S.F.)

* Correspondence: fanny.bonnet@ensc-lille.fr (F.B.); marc.visseaux@ensc-lille.fr (M.V.); Tel.: +33-320-434-091 (F.B.); +33-320-336-483 (M.V.)

Received: 15 November 2017; Accepted: 30 November 2017; Published: 5 December 2017

Abstract: This mini-review focuses on recent advances on the synthesis, structure, and characterization of allyl-based rare earth organometallic complexes, with emphasis on their ability to catalyze the polymerization of non-polar monomers such as conjugated dienes, styrene, and their related copolymerization.

Keywords: rare earths; allyl; conjugated dienes; stereoselective; polymerization catalysis

1. Introduction

Organometallic rare earths (REs) chemistry was first introduced at the beginning of the 20th century [1] but for some time it attracted little interest due to the lack of characterization—and, thus, understanding—of these highly-reactive compounds. However, since the 1970s, with the advent of modern techniques of analysis and synthesis, it became possible to better apprehend the structure and reactivity of organometallic RE complexes, which promoted the growth of research in this field of chemistry and its application to catalysis. Thenceforth, chemists realized that the thorough study of the reactivity of allyl derivatives of the rare earths, besides alkyl and hydride derivatives, could greatly benefit the comprehension of mechanisms involved in rare earth-catalyzed polymerization of olefins and conjugated dienes [2–6].

Rare earth-based allyl complexes are a relatively new area of organometallic chemistry of the RE elements and it was not until 1975 that the first series of rare earth complexes bearing an allyl ligand, $Cp_2RE(C_3H_5)$ (RE = Sm, Er, Ho and Cp = C_5H_5), was successfully synthesized by Tsutsui and Ely [7]. Thereafter, allylic-substituted rare earths, ranging from mono- to tetra-substituted allyl complexes, have been explored, and the group of Taube was probably the most successful in this area during the 1980s and 1990s [8]. In terms of reactivity, the allyl moiety is of specific interest because it makes it possible to carry out a certain number of elementary organometallic reactions, such as those involved in catalytic processes (insertion reactions [9], hydrogenolysis [10,11], hydrosilylation [12], alkyl exchange [13], etc.). Consequentially, the [RE-(allyl)] species has demonstrated its ability to catalyze polymerization reactions, with a particular behavior towards non-polar monomers (Scheme 1), some of which are highly stereo-selective [14–16]. The [RE-(allyl)] moiety has also been studied as a model for the chain initiation in olefin polymerization [17], and the coordination of the allyl ligand within the complexes, as well as the specificity of the rare earth metals used have shown to vary the outcome of the polymerizations.

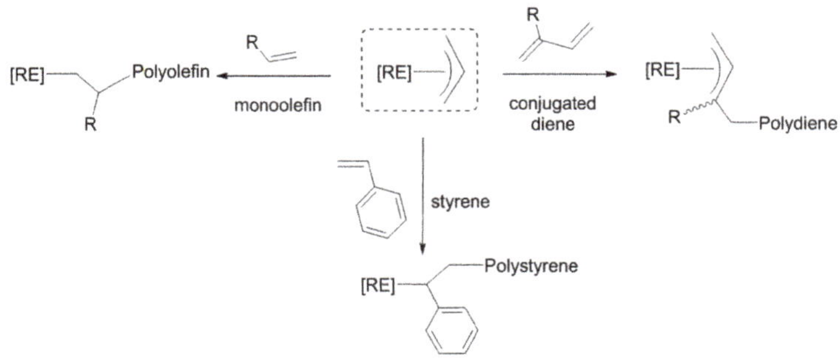

Scheme 1. Non-polar monomers polymerized by RE(allyl) compounds.

In 2010, Carpentier et al. [18] reviewed allyl rare earth complexes that had been studied over the past decades, along with their reactivity. Since then, a number of reports have been published that enlarge the knowledge on this particular class of compounds. In this review, we focus on the very recent development dealing with RE allyl compounds, highlighting their ability to catalyze the polymerization of non-polar monomers, such as butadiene, isoprene, styrene, and related co-polymerizations.

2. Allyl Complexes for the Polymerization of Butadiene

Polybutadiene (PB), arising from the polymerization of butadiene, a petro-sourced monomer, was first synthesized in the 20th century by using a sodium-based catalyst. The most notable advancement regarding the polymerization of dienes was the discovery of the Ziegler-Natta catalysts, which are still being used for industrial scale process for more than half a century, producing highly stereoregular *cis*- or *trans*-PB (Scheme 2) [19]. Rare earth complexes, and among them allyl-based catalysts, took their part in this context showing their ability to produce highly stereoregular PB with high activities, especially with the studies done by Taube's group [8,14,20].

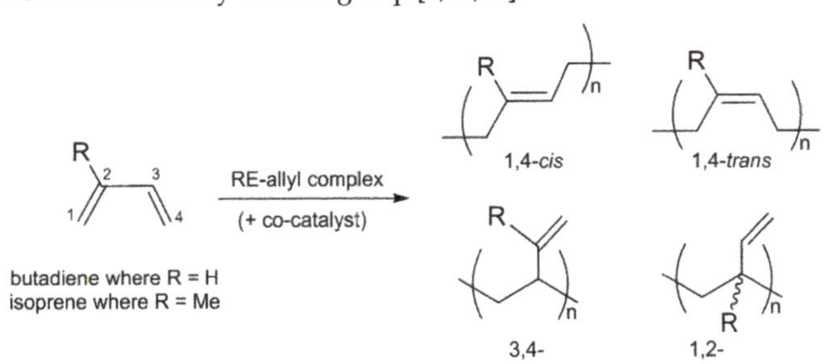

Scheme 2. Polymerization of butadiene and isoprene by means of RE(allyl)-based catalysts.

At the end of the 2000s, Okuda and co-workers reported the synthesis of a series of mono-cationic bis-allyl complexes $[RE(\eta^3\text{-}C_3H_5)_2(THF)_3]^+[B(C_6F_5)_4]^-$ (**1**$_{\mathbf{RE}}$, RE = Y, La, Nd, THF = tetrahydrofuran, Scheme 3) by reacting the tris-allyl complexes $RE(\eta^3\text{-}C_3H_5)_3$(dioxane) described by Taube with one equivalent of $\{[HNMe_2Ph][B(C_6F_5)_4]\}$ in THF [21,22]. The monocationic bis-allyl analogues bearing a non-perfluorinated counteranion $[RE(\eta^3\text{-}C_3H_5)_2(THF)_3]^+[BPh_4]^-$ (**1′**$_{\mathbf{RE}}$, RE = Y, La, Nd) were prepared similarly, but with $\{[HNEt_3Ph][BPh_4]\}$. In these complexes, the cationic allylic counterpart displays the same molecular structure as in complexes **1**$_{\mathbf{RE}}$. X-ray crystal study of the monocationic yttrium complex **1′**$_{\mathbf{Y}}$ showed that all three allyl ligands are η^3 coordinated with similar bond lengths between each allyl ligand and yttrium metal. Crystals of **1′**$_{\mathbf{La}}$ and **1′**$_{\mathbf{Nd}}$ were found to contain a fourth THF molecule, whereas elemental analysis was consistent with three-THF adducts after drying under vacuum. Two independent sets of ionic pairs were observed in **1′**$_{\mathbf{La}}$ and **1′**$_{\mathbf{Nd}}$, with allyl groups in a paddle-wheel fashion, or arranged as pincer-like towards each other. Shorter metal-allyl bonds were noticed by comparison with those of the parent compounds for both **1′**$_{\mathbf{La}}$ and **1′**$_{\mathbf{Nd}}$, revealing a higher Lewis acidity of the rare earth metal from neutral to cationic species. The NMR (Nuclear Magnetic Resonance) analysis of complexes **1**$_{\mathbf{Y}}$ and **1′**$_{\mathbf{Y}}$ displayed two signals for the allyl group corresponding to fast *syn/anti* exchange on the NMR timescale. In contrast, three distinct signals were seen for **1**$_{\mathbf{La}}$ (and **1′**$_{\mathbf{La}}$) and **1**$_{\mathbf{Nd}}$ (and **1′**$_{\mathbf{Nd}}$), typical of slow *syn/anti* exchange.

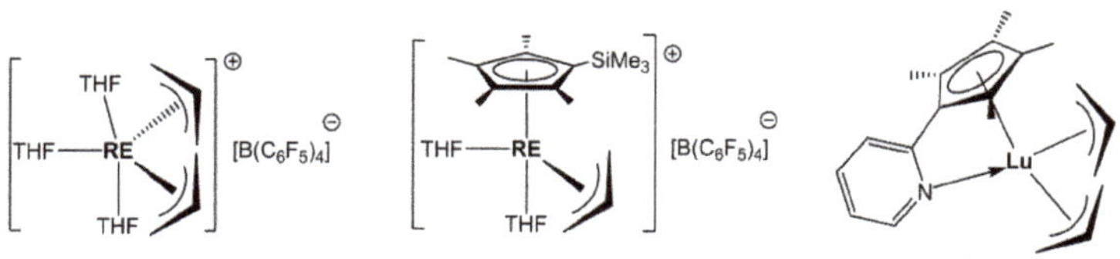

1$_{\mathbf{RE}}$, RE = Y, La, Nd, Sc **2**$_{\mathbf{RE}}$, RE = Y, La **3**$_{\mathbf{Lu}}$

Scheme 3. Allylic rare earth complexes used for butadiene polymerization [21,23,24].

The monocationic allyl complexes **1**$_{\mathbf{RE}}$ showed no activity towards the polymerization of 1,3-butadiene at room temperature [21]. In contrast, these complexes were active when combined with Al(^{i}Bu)$_3$ as co-catalyst. The catalyst made from the yttrium complex **1**$_{\mathbf{Y}}$ was found to be the most active (TOF 10,000 h^{-1}) leading to the formation of PB with the highest 1,4-*cis* stereoregularity of 90% in comparison to those made from **1**$_{\mathbf{La}}$ and **1**$_{\mathbf{Nd}}$ complexes (low yields and 33% and 75% of 1,4-*cis* units, respectively). This is a rare example that contradicts the well-known "neodymium effect" [6]. The in situ addition of one extra equivalent of {[NPhMe$_2$H][B(C$_6$F$_5$)$_4$]} to the former mono-cationic system in the polymerization mixture led to an increase in both the activity (up to TOF (Turn-Over Frequency) 12,000 h^{-1}) and the selectivity with 92.5% of 1,4-*cis*-PB in the case of **1**$_{\mathbf{Y}}$. However, the polymerizations carried out with **1**$_{\mathbf{Y}}$ as precatalyst gave much broader dispersity, i.e., a less controlled process. Interestingly, the lanthanum-based precatalyst **1**$_{\mathbf{La}}$ showed reverse stereo-selectivity when used with Al(^{i}Bu)$_3$ only and with the dual Al(^{i}Bu)$_3$/{[NPhMe$_2$H][B(C$_6$F$_5$)$_4$]} combination, switching from 63.3% 1,4-*trans* to 80.5% 1,4-*cis*-selectivity, respectively. The in situ formation of monoallyl dicationic species, as proposed by the authors, was likely to be responsible for the better reactivity. However, isolated [RE(η^3-C$_3$H$_5$)(THF)$_6$]$^{2+}${[B(C$_6$F$_5$)$_4$]$^-$}$_2$ (RE = La, Nd) from bulk scale syntheses were found to be rather unreactive.

The scandium congener [Sc(η^3-C$_3$H$_5$)$_2$(THF)$_3$]$^+$[B(C$_6$F$_5$)$_4$]$^-$ (**1**$_{\mathbf{Sc}}$) was prepared similarly as **1**$_{\mathbf{RE}}$ (RE = Y, La, Nd) from the newly synthesized tris-allyl scandium, which was lacking in the family of analogous complexes of rare earths until Okuda and coworkers succeeded to isolate it [23]. Complex **1**$_{\mathbf{Sc}}$ could also be synthesized by reacting K[Sc(C$_3$H$_5$)$_4$] with 2 equiv. {[HNEt$_3$][BPh$_4$]}. ^{1}H NMR analysis of **1**$_{\mathbf{Sc}}$ revealed dynamic behavior of the allyl group while the crystallographic distances were typical of the η^3 coordination mode. This scandium allyl complex was assessed towards polymerization of 1,3-butadiene. When it was combined with 1 equiv. of Al(^{i}Bu)$_3$ as co-catalyst under similar conditions as for **1**$_{\mathbf{RE}}$ (RE = Y, La, Nd), it gave rise to a little amount of PB (7% yield, TOF 280 h^{-1}) with low stereoregularity (<60% 1,4-*cis*). The activity was improved (TOF 1150 h^{-1}) with the addition of one equivalent of [HNMe$_2$Ph][B(C$_6$F$_5$)$_4$] in the polymerization mixture, but no change in the selectivity was observed.

The same research group synthesized the half-sandwich mono-allyl complexes [RE(η^5-C$_5$Me$_4$SiMe$_3$)(η^3-C$_3$H$_5$)(THF)$_2$]$^+$[B(C$_6$F$_5$)$_4$]$^-$ (**2**$_{\mathbf{RE}}$, RE = Y, La, Scheme 3) by protonation of the bis-allyl complexes RE(η^5-C$_5$Me$_4$SiMe$_3$)(η^3-C$_3$H$_5$)$_2$(THF) with one equivalent of {[NPhMe$_2$H][B(C$_6$F$_5$)$_4$]} [21]. Little rigidity was noticed for the allyl group of **2**$_{\mathbf{La}}$ by ^{1}H NMR, whereas **2**$_{\mathbf{Y}}$ displayed higher fluxionality with fast *syn/anti* exchange. The activity of these mono-cationic complexes was screened towards the polymerization of 1,3-butadiene in the presence of Al(^{i}Bu)$_3$ (5 equiv.) as a co-catalyst, since the complexes were found to be inactive on their own. The lanthanum system showed lower activity (TOF 1600 h^{-1}), in comparison to the yttrium one which gave quantitative conversion for the same reaction time at room temperature in toluene (TOF 12,000 h^{-1}) and moderate selectivity (86% 1,4-*cis*).

Jian et al. prepared the pyridyl-functionalized half-sandwich of lutetium, (C_5Me_4-C_5H_4N) Lu(η^3-C_3H_5)$_2$ $\mathbf{3_{Lu}}$ by the clean protonolysis reaction of Lu(η^3-C_3H_5)$_3$(dioxane) with C_5Me_4H-C_5H_4N in THF [24]. The X-ray analysis showed that both allyl groups coordinate to the lutetium in a η^3 mode. Due to the coordination of the pyridyl moiety, the complex was isolated as solvent free. ^{1}H NMR displayed the typical 1 (quintet)/4 (doublet) set of signals for allyl groups in dynamic equilibrium. In combination with trityl borate activator, highly active catalyst towards butadiene polymerization was formed (TOF 60,000 h^{-1}), which was also *cis*-1,4-selective up to 97%. When chlorobenzene was used as the solvent, a drop in catalytic activity was noticed, but with the benefit of the stereo-conversion of *cis*-1,4 PB (99%).

3. Allyl Complexes for the Polymerization of Isoprene

The coordination polymerization of isoprene can lead to the formation of polyisoprene (PI) containing four different isomers: 1,4-*cis*, 1,4-*trans*, 3,4-, and 1,2- (Scheme 2). Natural rubber extracted from the *Hevea* tree is composed predominantly of *cis*- units, whereas the one extracted from Gutta Percha is mainly *trans*-. Both *cis*-and *trans*-PI found numerous applications in the fields of adhesives, sports equipment, or the tire industry [25].

Cui, Hou, and co-workers explored constrained-geometry-conformation (CGC) allyl complexes of the rare earths as catalysts for isoprene polymerization [26]. They synthesized the aminophenyl-cyclopentadienyl complexes (C_5Me_4–C_6H_4-*o*-NMe_2)RE(η^3-C_3H_5)$_2$ ($\mathbf{4_{RE}}$, RE = Y, Nd, Gd, Dy, Scheme 4) by first reacting (C_5Me_4–C_6H_4-*o*-NMe_2)Li with $RECl_3(THF)_n$ and then adding the Grignard reagent, C_3H_5MgCl. By ^{1}H NMR, the allyl group in $\mathbf{4_Y}$ was found fluxional with the typical 1H/4H resonances. The yttrium, gadolinium, and dysprosium complexes were characterized by X-ray crystallography as solvent-free—even though the reaction was performed in THF solvent—isostructural complexes and it showed that both allyl moieties coordinate in the η^3 mode. The activity of these bis(allyl) complexes was assessed towards the polymerization of isoprene in the presence of AlR_3 (mainly Al(^{i}Bu)$_3$) and {[PhMe$_2$NH][B(C_6F_5)$_4$]} in toluene at 20 °C. The neodymium complex $\mathbf{4_{Nd}}$ had the highest activity (TOF 3000 h^{-1}), followed by the gadolinium $\mathbf{4_{Gd}}$ (1000 h^{-1}) and then the dysprosium $\mathbf{4_{Dy}}$ analogue (110 h^{-1}). The yttrium complex $\mathbf{4_Y}$ was almost inert towards this polymerization under these conditions, while low to medium activity (TOF 140–800 h^{-1}) was noticed with {[Ph$_3$C][B(C_6F_5)$_4$]}/AlR_3 (R = Me, Et, ^{i}Bu). The gadolinium complex $\mathbf{4_{Gd}}$ afforded the highest *cis*-regular PI at 99.2% (at 0 °C), along with the living character of the polymerization. It was observed that when the Al/Gd ratio was increased, a typical catalyzed chain growth (CCG) process was operating, with regular decrease of the molecular weight of the PI while the molecular weight distribution remained unchanged. Al(^{i}Bu)$_3$ behaved partially as a chain transfer agent, with ca. eight growing PI chains per RE metal, while no drop of stereo-selectivity was evident (ca. 98% *cis*-units).

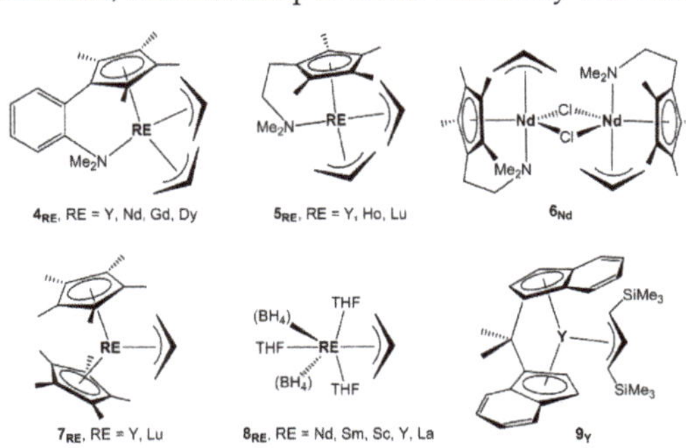

Scheme 4. Allylic rare earth complexes used for isoprene polymerization [26–31].

Jende et al. synthesized the allyl half-sandwich complexes of small-size rare earths, $Cp^{NMe2}RE(\eta^3\text{-}C_3H_5)_2$ (Cp^{NMe2} = $C_5Me_4CH_2CH_2NMe_2$; $\mathbf{5_{RE}}$, RE = Y, Ho, Lu, Scheme 4) having a more flexible (*N,N*-dimethylamino)ethyl-functionalized cyclopentadienyl ligand than in the $\mathbf{4_{RE}}$ complexes [27]. The reaction was conducted in two steps by first reacting $RECl_3(THF)_x$ with $Cp^{NMe2}Li$ at room temperature, and then adding two equivalents of the Grignard reagent C_3H_5MgCl. X-ray analysis showed that the $\mathbf{5_{RE}}$ complexes were under a bis(allyl) half-sandwich monomeric form, and were all isostructural. One of the allyl groups showed similar bond lengths between terminal and central carbon atoms, while the second allyl group showed a significantly longer bond length between the rare earth metal and the terminal carbon. The overall moiety arrangement was likened to that of $\mathbf{4_Y}$ [26]. The ^1H NMR of both diamagnetic complexes $\mathbf{5_Y}$ and $\mathbf{5_{Lu}}$ showed a similar quintet (1H)/doublet (4H) set of signals characteristic of dynamic exchange of the allylic protons, along with splitting of the quintet in the case of $\mathbf{5_Y}$ being attributable to the coupling with ^{89}Y. When the same synthetic procedure as for $\mathbf{5_{RE}}$ was done in the case of the larger size neodymium element, it gave a monoallyl chloro derivative $[Cp^{NMe2}Nd(\eta^3\text{-}C_3H_5)(\mu\text{-}Cl)]_2$ ($\mathbf{6_{Nd}}$, Scheme 4) instead of the expected bis allyl half-sandwich. Complex $\mathbf{6_{Nd}}$ was found to be dimeric through (μ-Cl) bridges with one substantially longer Nd-Cl bond than the other one, anticipating a possible reactivity. [Allyl]/[Cl] exchange was observed when complexes $\mathbf{5_Y}$ and $\mathbf{5_{Ho}}$ were reacted with $AlEt_2Cl$, affording multi(μ-chlorido) hexametallic $[RE_6Cl_{12}]$ clusters. This was in agreement with the observation that the combination of any $\mathbf{5_{RE}}$ with $AlEt_2Cl$ was found inert towards isoprene polymerization.

When activated with either $\{[Ph_3C][B(C_6F_5)_4]\}$ or $\{[PhNMe_2H][B(C_6F_5)_4]\}$ borates, the half-sandwiches $\mathbf{5_Y}$ and $\mathbf{5_{Ho}}$ were found to be poorly active (TOF 100 h^{-1}) towards the polymerization of isoprene, while $\mathbf{5_{Lu}}$ displayed higher activity (TOF 500 h^{-1}). The yttrium ($\mathbf{5_Y}$) and holmium ($\mathbf{5_{Ho}}$) complexes afforded predominantly 3,4-PI with both co-catalysts. The lutetium complex $\mathbf{5_{Lu}}$ afforded non stereo-regular PI. The dispersities were very narrow (1.04–1.17) for all the precatalysts used, accounting for unique active species. In addition, in the presence of 10 equiv. $AlMe_3$, the $\mathbf{5_Y}$/borate and $\mathbf{5_{Ho}}$/borate combinations afforded much more active catalysts (TOF up to 2000 h^{-1}) with a switch in selectivity towards *trans*-1,4 selectivity (71% Y, 72% Ho). In contrast, when 10 equiv. Al^iBu_3 were added to the $\mathbf{5_Y}$/borate and $\mathbf{5_{Ho}}$/borate systems, it afforded a major selectivity towards *cis*-1,4 PI (74% Y, 74% Ho) along with improved activity (TOF 1000 h^{-1}). For the lutetium complex $\mathbf{5_{Lu}}$, when $AlMe_3$ or $Al(^iBu)_3$ was added to the system the main effect was a similar gain of activity (TOF up to 2000 h^{-1}), but with no improvement of the stereocontrol. A decrease in the polymer molecular weights with narrow distributions was also noticed, which indicated a chain transfer to aluminum. The activity of the chloroallyl neodymium complex $\mathbf{6_{Nd}}$ was also assessed towards the polymerization of isoprene with the use of either $\{[Ph_3C][B(C_6F_5)_4]\}$ or $\{[PhNMe_2H][B(C_6F_5)_4]\}$ as a borate activator. There was no activity when the [Nd]:[borate] ratio was 1:1. However, when 1 equiv. of activator was added to the dinuclear complex ([Nd]:[borate] is 2:1), PI, with mainly 3,4-motives of up to 66%, was isolated with good activity (TOF 500 h^{-1}), along with narrow dispersities (1.10–1.11). When 10 equiv. of $AlMe_3$ was added to the system along with $\{[PhNMe_2H][B(C_6F_5)_4]\}$, there was a switch in stereo-selectivity and *trans*-1,4-PI (85%) was obtained, while the addition of $Al(^iBu)_3$ (10 equiv.) gave rise to 3,4-PI (85%). The activity was improved by a factor of four (TOF 2000 h^{-1}) and of two (TOF 1000 h^{-1}) by the addition of $AlMe_3$ and $AliBu_3$, respectively. In all cases with complexes bearing this Cp^{NMe2} ligand ($\mathbf{5_{RE}}$ and $\mathbf{6_{Nd}}$), the use of $AliBu_3$ vs. $AlMe_3$ provided reversible transfer between the RE metal and the aluminum during the polymer chain growing process. Allyl(RE)-alkyl(Al) exchange was evidenced by ^1H NMR experiments to support the formation of the polymerization active species. None of these allyl complexes $\mathbf{5_{RE}}$ and $\mathbf{6_{Nd}}$ were found active on their own without activator/co-catalyst.

The photopolymerization of isoprene mediated by $(C_5Me_5)_2RE(\eta^3\text{-}C_3H_5)$ ($\mathbf{7_{RE}}$, RE = Y, Lu, Scheme 4) was assessed [28]. These complexes had been previously synthesized and characterized [13]. In particular, the ^1H NMR analysis established non-fluxional allyl group with 1H/2H/2H allyl signals. In neat monomer, the reaction resulted in obtaining low molecular weight PI with 3,4-units being slightly predominant, which was consistent with radical polymerization, according to the authors. In the absence of irradiation, no polymerization was observed with the yttrium complex $\mathbf{7_Y}$.

Bonnet, Visseaux, and co-workers synthesized the first RE complexes bearing both allyl and borohydride ligands, $RE(BH_4)_2(\eta^3\text{-}C_3H_5)(THF)_3$ ($\mathbf{8_{RE}}$, RE = Nd, Sm, Scheme 4) by reacting $RE(BH_4)_3(THF)_3$ with half an equivalent of $Mg(C_3H_5)_2(L)_n$ (L = THF, dioxane) in THF at room temperature [29]. From ^{1}H NMR analysis, the allyl moiety appeared as a 1H/2H/2H set of resonances for both $\mathbf{8_{RE}}$ complexes, revealing no dynamic behavior at the ^{1}H NMR timescale. X-ray analyses showed that both complexes were monomeric and isostructural, with the two borohydride ligands being tridentate. The activity of these mixed borohydrido-allyl rare earth complexes $RE(BH_4)_2(\eta^3\text{-}C_3H_5)(THF)_3$ was assessed towards the polymerization of isoprene. Whereas the samarium complex showed no reactivity, its neodymium analogue was found to be active, either on its own, due to the presence of the Nd-allyl bond, or combined with various alkylating reagents. When the latter was tested alone, highly *trans*-regular PI with 92.2% *trans*-selectivity along with Đ = 1.54 was obtained with moderate activity (TOF 177 h^{-1}). In the presence of one equivalent of $Mg(^n Bu)(Et)$ with respect to $\mathbf{8_{Nd}}$, the activity was substantially increased (TOF 425 h^{-1}), but also the *trans*-selectivity of the reaction was improved (95.5% *trans*). With aluminum-based co-catalysts, such as $Al(^i Bu)_3$ or MAO (MethylAlumOxane), the activities were greatly improved with TOF of 1000 h^{-1}, however, the *trans*-selectivity was affected (78.7% and 68.2%, respectively). This family of complexes was recently extended to scandium (bis-THF adduct), yttrium, and lanthanum [30].

The *ansa*-lanthanidocene allyl *rac*-$\{Me_2C(Ind)_2\}Y[\eta^3\text{-}1,3\text{-}(SiMe_3)_2C_3H_3]$ ($\mathbf{9_Y}$) (Ind = 2-indenyl) was assessed for the polymerization of isoprene [32]. This compound was synthesized previously and initially evaluated for styrene polymerization as single-component catalyst [31]. Towards isoprene, and again in the absence of co-reagent, complex $\mathbf{9_Y}$ afforded 1,4-*trans* PI (87–91%) with moderate activity (TOF 70 h^{-1}). It is noteworthy that, up to now, this is the unique example of a single-component yttrium catalyst for the *trans*-stereo-selective polymerization of isoprene. With di(ethyl)zinc in excess, it was established that reversible Y/Zn chain transfer was operating, with a comparable activity (TOF 76 h^{-1}), while maintaining the 1,4-*trans* selectivity (ca. 90%). In turn, the polymerization of isoprene mediated by $\mathbf{9_Y}/Mg(^n Bu)_2$ occurred with a good level of transfer, but at the expense of the 1,4-*trans* selectivity (up to 47% 3,4 units).

4. Allyl Complexes for the Polymerization of Styrene

Polystyrene (PS) is a thermoplastic polymer mostly known for its applications in long-lasting packaging. Styrene, when polymerized by coordination-insertion polymerization, can give rise to PS under three different forms: isotactic, syndiotactic, or atactic (Scheme 5).

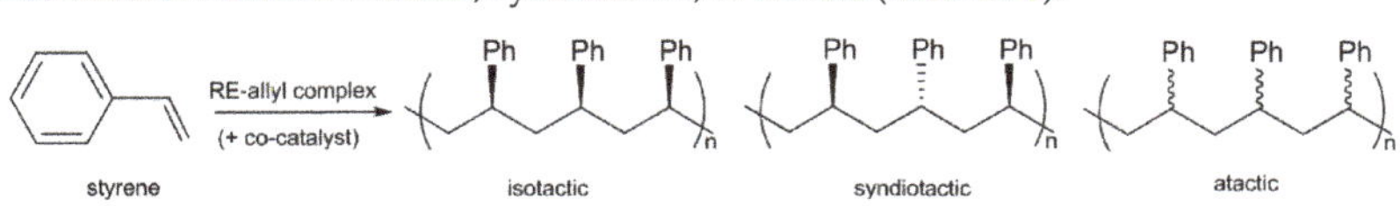

Scheme 5. Polymerization of styrene by means of RE(allyl)-based catalysts.

The scandium bis-allyl mono-cationic complex $[Sc(\eta^3\text{-}C_3H_5)_2(THF)_3]^+[B(C_6F_5)_4]^-$ ($\mathbf{1_{Sc}}$), which was proved to be efficient towards butadiene polymerization when combined with $Al(^i Bu)_3$ and $\{[HNMe_2Ph][B(C_6F_5)_4]\}$ (see above) was also studied for styrene polymerization. It was found to be inactive in this case as a single component, while displaying low activity in the presence of $Al(^i Bu)_3$ (TOF up to 95 h^{-1}) to afford atactic PS. The addition of $\{[HNMe_2Ph][B(C_6F_5)_4]\}$ as activator did not really improve the catalysis (atactic PS, TOF 114 h^{-1}) [23].

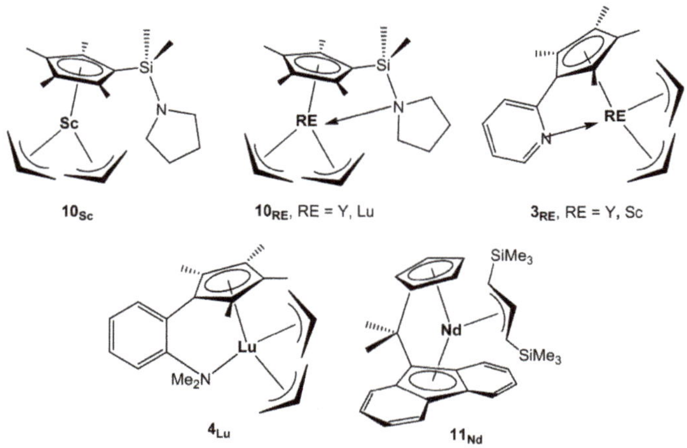

Scheme 6. Allylic rare earth complexes used for styrene polymerization [33–35].

The pyrrolidinyl-functionalized half-sandwich complexes $(C_5Me_4SiMe_2NC_4H_8)RE(\eta^3\text{-}C_3H_5)_2$ (**10_{RE}**, RE = Sc, Y, Lu, Scheme 6) were synthesized by reacting $RECl_3$ with one equivalent of $C_5Me_4SiMe_2\text{-}NC_4H_8Li$ followed by the addition of two equivalents of C_3H_5MgCl in THF at room temperature [33]. The ^1H NMR spectra of the three complexes indicated the fluxional allyl ligand in solution, with one sharp doublet signal for the terminal allylic protons and one multiplet for the central allylic protons. X-ray analysis showed that for the scandium complex **10_{Sc}**, the pendant pyrrolidinyl ligand does not coordinate to the metal center through the nitrogen, whereas this coordination was present in the yttrium **10_Y** and lutetium **10_{Lu}** complexes due to higher size of the RE^{3+} cation [36]. In **10_{Sc}**, the two allyl moieties were coordinated to the central metal in η^3 mode with one allyl group prone and the other supine. The half sandwich complex **10_{Sc}** was found to be highly active towards styrene polymerization when activated with one equivalent of $\{[Ph_3C][B(C_6F_5)_4]\}$ in toluene at room temperature, producing pure syndiotactic PS (TOF 250 h^{-1}). Yttrium complex **10_Y** was much less active (70 h^{-1}) while **10_{Lu}** only produced traces of the polymer under the same conditions. When an excess of $Al(^iBu)_3$ was added to **10_{Sc}**/$\{[Ph_3C][B(C_6F_5)_4]\}$, the activity increased drastically (TOF 1500 h^{-1}).

The pyridyl-functionalized half-sandwich complexes $(C_5Me_4\text{-}C_5H_4N)RE(\eta^3\text{-}C_3H_5)_2$ (**3_{RE}**, RE = Y, Sc, Scheme 6), analogs of complex **3_{Lu}**, were prepared. The syntheses were conducted by metathetic reaction of $(C_5Me_4\text{-}C_5H_4N)Li$ with 1 equiv. $RECl_3$, followed by addition of 2 equiv. allyl MgCl in THF at room temperature [35]. Alternatively, the target compounds could be also obtained by the acid base reaction between $RE(C_3H_5)_3$(1,4-dioxane) and the pyridyl-cyclopentadiene $C_5Me_4H\text{-}C_5H_4N$, as same as previously done for **3_{Lu}**. The X-ray structure displayed the expected CGC-geometry with η^5(Cp)/κ^1(Py) coordination to the RE center. In contrast to what was observed in complexes **2_Y** [21] and **4_Y** [26], the ^1H NMR of **3_Y** displayed a 1H/2H/2H pattern resonances for the allyl ligands, thus suggesting poor fluxionality in solution, while in the case of **3_{Sc}** only two allyl signals (1H/4H) were observed. In combination with $\{[Ph_3C][B(C_6F_5)_4]\}$ in toluene, complex **3_Y** showed moderate activity (TOF 120 h^{-1}) but syndiotactic enriched PS (*rrrr* = 88%). This catalytic system made of yttrium was much more active in chlorobenzene (TOF 2000 h^{-1}) but less stereoselective. In turn, the scandium analogue **3_{Sc}** was found to display under the same conditions exceptionally high performances affording PS perfectly syndiotactic (*rrrr* > 99%) with TOF value of 60,000 h^{-1} (complete conversion of 1000 equivalents of monomer in 1 min at 20 °C in toluene) [37] and narrow dispersity (Đ = 1.40–1.50). However, lower activity/selectivity was noticed in chlorobenzene for **3_{Sc}**. Although the process was less controlled (Đ = 1.94), the lutetium complex **3_{Lu}** gave the same notable results as **3_{Sc}** in terms of catalytic capability, and remarkably, the former complex exhibited a rare dual catalysis ability [24,38] in both syndiotactic styrene polymerization and *cis*-selective butadiene polymerization. Contrary to what advanced for $\mathbf{4_{Lu}}$, the smaller Cp_{cent}-RE-N bite angle in pyridyl-Cp complexes $\mathbf{3_{RE}}$, along with a more electron withdrawing effect of the ligand, was proposed to explain the catalytic efficiency of the latter complexes.

Cui, Hou, and co-workers extended the family of constrained geometry catalysts $\mathbf{4_{RE}}$ [26] to the lutetium derivative (C_5Me_4–C_6H_4-*o*-NMe_2)Lu(η^3-C_3H_5)$_2$ ($\mathbf{4_{Lu}}$, Scheme 6), by a synthetic procedure similar to that used for the latter complexes [35]. Upon activation of this lutetium complex with {[Ph_3C][B(C_6F_5)$_4$]}, or with the {[Ph_3C][B(C_6F_5)$_4$]}/Al(^{i}Bu)$_3$ combination, the product was inert for the polymerization of styrene. The same was observed for $\mathbf{4_Y}$ under the same conditions. According to the authors, this could be due to the large Cp_{cent}-Lu-N bite angle in $\mathbf{4_{RE}}$, in comparison with the value determined in complexes $\mathbf{3_{RE}}$, which hinders the coordination and insertion of the styrene monomer.

Carpentier and colleagues found that the combination of bulky allyl *ansa*-lanthanidocenes *rac*-{$Me_2C(Ind)_2$}Y[η^3-1,3-($SiMe_3$)$_2C_3H_3$] ($\mathbf{9_Y}$) and {Me_2C(Cp)(Flu)}Nd[η^3-1,3-($SiMe_3$)$_2C_3H_3$] ($\mathbf{11_{Nd}}$) (Flu = 9-fluorenyl, Scheme 6) with di(n-butyl)magnesium in excess behaved efficiently as binary catalytic systems for the stereo-controlled coordinative polymerization of styrene under reversible chain transfer regime (CCTP, coordinative chain transfer polymerization). Isotactic PS was produced with $\mathbf{9_Y}$/Mg(^{n}Bu)$_2$ while $11_{\mathbf{Nd}}$/Mg(^{n}Bu)$_2$ yielded syndiotactic PS, both with high activities (TOF up to 2100 h^{-1} and 2500 h^{-1}, respectively [34]. By adjusting the amount of Mg(^{n}Bu)$_2$, up to 200 polymer chains can be generated per RE center. Complex $\mathbf{9_Y}$ was previously shown to be active as a single-component catalyst [32], but the dispersity was improved in the presence of excess Mg(^{n}Bu)$_2$. Mechanistic investigations, also confirmed by the support of theoretical studies, demonstrated that the initiation of the polymerization resulted from the insertion of styrene into the RE-allyl (single component) or RE-alkyl (chain transfer) moiety, and that an enantiomorphic site control mechanism (ECM) was operative to account for the isoselectivity observed [39]. As for the *ansa* derivative $\mathbf{11_{Nd}}$, it was synthesized by ionic metathesis between K[1,3-(SiMe3)2C3H3] and [{Me_2C(Cp)(Flu)}Nd(μ-Cl)]$_2$ and was found unsolvated. It was shown that $\mathbf{11_{Nd}}$ acts as a single-component catalyst for the polymerization of styrene, and produces *s*PS albeit at a much lower rate (TOF = 20–60 h^{-1}) than the regular allyl compounds {Me_2C(Cp)(Flu)}RE(η^3-C_3H_5)(THF) (RE = Y, La, Nd, 1000–17,000 h^{-1}) [40]. Using DFT (Density Functional Theory) studies, the origin of the syndiospecificity control, due to a chain-end control mechanism (CEM), was proposed to result from the conjunction of the minimization of two repulsion effects: the classical phenyl (incoming monomer)-phenyl (last unit inserted) one during the growing of the polymer chain, and also of the repulsion between the fluorenyl ligand and the incoming styrene unit [41].

5. Allyl Complexes for Co-Polymerization

Since 2010, very few reports dealt with the copolymerization of dienes/styrene monomers involving allyl derivatives of the rare earths.

Complex $\mathbf{3_{Lu}}$, which was found efficient as a precatalyst towards butadiene and styrene homopolymerization (see above), was assessed successfully for the copolymerization of these two monomers. When combined with trityl borate, the resulting catalyst system showed high activity (TOF up to 5100 h^{-1}) for the statistical and the sequenced copolymerization of butadiene and styrene. The concurrent addition of both the monomers with the styrene feed molar fractions ranging from 10–90%, afforded highly *cis*-1,4-regulated PB and syndiotactic PS segments, high molecular weight (M_n = 8.8–12.1 × 10^4 g mol^{-1}) and narrow dispersity (1.29–1.68) across the styrene feed ratio range. Kinetic studies demonstrated that in the presence of a mixture of the two monomers butadiene was consumed first, followed by the growing of the syndiotactic PS sequence, finally affording diblock styrene-butadiene copolymers in all cases. The same kind of copolymer was isolated from sequenced butadiene/styrene copolymerization. These copolymers displayed a phase separated morphology of the hard (*s*PS) and soft (*cis*-PBu) domains, as observed by atomic force microscopy [24].

Upon activation with {[Ph_3C][B(C_6F_5)$_4$]}/Al(^{i}Bu)$_3$, (C_5Me_4–C_6H_4-*o*-NMe_2)RE(η^3-C_3H_5)$_2$ ($\mathbf{4_{Gd}}$) afforded via sequential monomer addition unprecedented *cis*-PI-*b*-*cis*-PB and *cis*-PI-*b*-*cis*-PB-*b*-*cis*-PI block copolymers. This was possible thanks to the living polymerization process occurring through reversible Gd/Al chain transfer [26].

The allyl *ansa* complex **9**$_{Y}$ showed unique behavior to produce isoprene-styrene copolymers. Polymerizations experiments were conducted in one pot in the absence of co-reagent and afforded copolymers with blocky distribution of the two monomers. Unprecedented well-defined and crystalline 1,4-*trans*-PI-*b*-iPS diblock copolymers were also prepared by sequential addition of the two monomers [31].

6. Concluding Remarks

In the last eight years, a number of new allyl rare earth complexes have been synthesized and assessed towards the (co-)polymerization of non-polar monomers (Table 1). It must be noted that when browsing through the recent reports of the field gathered in this review, it appears that the utilization of such compounds for polymerization is limited to conjugated dienes and styrene, while none deal with ethylene, although allyl species of the rare earths are known to mediate the polymerization of that latter monomer [3,6].

In some cases towards conjugated dienes or styrene, an allyl rare earth complex is active by itself, i.e., no co-catalyst is necessary to initiate the polymerization. However, this is limited to the case of neodymium, or bulky *ansa*-metallocenes. In general, the performances are improved when an alkylating reagent and/or an activator are associated to the allylic compound. Regarding conjugated dienes, allyl complexes afford catalysts that enable the production of polymers with high stereoselectivity, along with very high activities.

The complexes synthesized by most research teams often focus on the "small" REs, i.e., the *late* lanthanides—having small size ionic radius—also including yttrium and scandium, especially when they are of the CGC-type. In turn, when it comes to metallocene-like derivatives, "big" RE, i.e., the *early* lanthanides, also including lanthanum, are privileged. In general, many scandium complexes in a given series afford the best catalyst, particularly as far as pseudo-cationic processes are concerned, which corroborates a recent theoretical study by Hou and coworkers [42].

Most allyl complexes of the RE described in this review are cyclopentadienyl derivatives, at the exception of the mixed allyl-borohydrides series recently reported. Such mixed allyl-borohydrides undoubtedly foreshadow a new platform towards novel families of allyl rare earth post-metallocenes.

Although allyl complexes of the rare earths may sometimes be difficult to isolate, their synthesis is clearly worth the effort. Indeed, the hapticity of the allyl ligand can assist in isolating a compound where the alkyl analog is not stable, and also limits the coordination of an additional solvent molecule, which may be detrimental to the catalytic performances. Moreover, as illustrated in this mini-review, allyl complexes may be obtained under a monomeric form, which can favor the reactivity vs. bridged alkyl complexes. So far, the presence of an allyl ligand does not guarantee the reactivity towards the insertion reaction of a monomer molecule and, hence, the polymerization: this allyl moiety must be effectively reactive. The fluxionality of the allyl ligand, as seen by proton NMR, can be an indication of a possible reactivity: it is quite frequently observed that the most active catalyst in a series often corresponds to the complex whose allyl ligand has a certain degree of fluxionality. However, this does not seem to be generalizable for the larger rare earths.

Nevertheless, in most cases the help of a co-catalyst and/or an activator is mandatory, at least to improve the catalytic performances, or to better control the process, especially when the reactions are conducted under reversible chain transfer conditions.

It can be anticipated that further research will be conducted in this area of chemistry for the years to come.

Table 1. Summary of complexes and their polymerizations.

Complex [1]	Allyl Flux. (^1H NMR)	Cocat.	Activ. [2]	Butadiene [1] Sel./TOF [3]	Isoprene [1] Sel./TOF [3]	Styrene [1] Sel./TOF [3]	Copolymers [1] Sel./TOF [3]	Reference
$\mathbf{1}_{\underline{Y}}$, $\mathbf{1}_{La}$, $\mathbf{1}_{Nd}$	yes (Y) no (La, Nd)	-	-	inactive				[13]
		Al^iBu_3	-	1,4-*cis* 90%/10,000 h^{-1}				
			HNB	1,4-*cis* 92.5%/12,000 h^{-1}				
$\mathbf{1}_{Sc}$	yes	Al^iBu_3	-	not selective/280 h^{-1}				[23]
			HNB	not selective/1600 h^{-1}				
		-	-	inactive		inactive		
		Al^iBu_3	-			atactic/95 h^{-1}		
			HNB			atactic/114 h^{-1}		
$\mathbf{2}_{\underline{Y}}$, $\mathbf{2}_{La}$	yes (Y) no (La)	-	-	inactive				[21]
		Al^iBu_3		1,4-*cis* 86%/12,000 h^{-1}				
$\mathbf{3}_{Lu}$	yes	-	TB	1,4-*cis* 97%/60,000 h^{-1}		Syndiotactic/60,000 h^{-1}	*cis*-PI-*b*-sPS/5000 h^{-1}	[24]
$\mathbf{3}_{Y}$, $\mathbf{3}_{\underline{Sc}}$	no (Y) yes (Sc)	-	TB			Syndiotactic/60,000 h^{-1}		[35]
$\mathbf{4}_{Y}$, $\mathbf{4}_{Nd}$, $\mathbf{4}_{\underline{Gd}}$, $\mathbf{4}_{Dy}$	yes (Y)	Al^iBu_3	HNB		1,4-*cis* 99%/1000 h^{-1} (3000 h^{-1}, Nd) CCG with excess Al^iBu_3		*cis*-PI-*b*-*cis*-PB; *cis*-PI-*b*-*cis*-PB-*b*-*cis*-PI	[26]
$\mathbf{4}_{Y}$, $\mathbf{4}_{Lu}$		-	HNB			inactive		
		Al^iBu_3	HNB			inactive		
$\mathbf{5}_{\underline{Y}}$, $\mathbf{5}_{Ho}$, $\mathbf{5}_{Lu}$	yes (Y, Lu)	-	-		inactive			[27]
		-	HNB/TB		3,4- 79%/100 h^{-1} (500 h^{-1}, Lu)			
$\mathbf{5}_{Y}$, $\mathbf{5}_{\underline{Ho}}$, $\mathbf{5}_{Lu}$		$AlMe_3$	HNB		1,4-*trans* 72%/2000 h^{-1}			
		Al^iBu_3			1,4-*cis* 74%/1000 h^{-1}			
$\mathbf{6}_{Nd}$		-	-		inactive			[27]
		-	HNB/TB		3,4- 66%/500 h^{-1}			
		$AlMe_3$	HNB		1,4-*trans* 85%/2000 h^{-1}			
		Al^iBu_3			3,4- 85%/1000 h^{-1}			

Table 1. *Cont.*

Complex [1]	Allyl Flux. (^{1}H NMR)	Cocat.	Activ. [2]	Butadiene [1] Sel./TOF [3]	Isoprene [1] Sel./TOF [3]	Styrene [1] Sel./TOF [3]	Copolymers [1] Sel./TOF [3]	Reference
7_{Y}, 7_{Lu}	no (Y, Lu) [23]	-	photo-activation		not selective (3,4 major)			[28]
8_{Nd}, 8_{Sm}	no (Nd, Sm)	-	-		1,4-*trans* 92%/180 h^{-1}			[29]
		$Mg^{n}Bu_2$	-		1,4-*trans* 95.5%/425 h^{-1}			
		$Al^{i}Bu_3$	-		1,4-*cis* 78%/1000 h^{-1}			
9_{Y}		-	-		1,4-*trans* 91%/70 h^{-1}	Isotactic/14,400 h^{-1} [31]	1,4-*trans*-PI-*co*-iPS 1,4-*trans*-PI-*b*-iPS	[32]
		$Mg^{n}Bu_2$	-		1,4-*trans* 90%/76 h^{-1} CCTP with excess $Mg^{n}Bu_2$			
		$Mg^{n}Bu_2$	-			Isotactic/2100 h^{-1} CCTP with excess $Mg^{n}Bu_2$		[34]
<u>10_{Sc}</u>, **10_Y**, **10_{Lu}**	yes (Sc, Y, Lu)	-	TB			Syndiotactic/250 h^{-1}		[35]
		$Al^{i}Bu_3$	TB			Syndiotactic/1500 h^{-1}		
11_{Nd}		-	-			Syndiotactic/60 h^{-1}		[34]
		$Mg^{n}Bu_2$	-			Syndiotactic/2500 h^{-1} CCTP with excess $Mg^{n}Bu_2$		

[1] The activity/selectivity refers to the best result obtained for a given metal within a given series, and corresponding to the underlined complex. [2] HNB = {[$HNMe_2Ph$][$B(C_6F_5)_4$]}, TB = {[CPh_3][$B(C_6F_5)_4$]}. [3] In mol monomer/mol catalyst/h.

Acknowledgments: The Ministère de l'Enseignement Supérieur et de la Recherche (Ph.D. grant to S.F.) the Région Nord-Pas de Calais, and the FEDER are acknowledged for supporting and funding this work. Y. Champouret is greatly acknowledged for careful reading of this manuscript.

Author Contributions: The authors contributed equally to the writing of the paper.

Conflicts of Interest: The authors declare no conflict of interest. The founding sponsors had no role in the design of the study; in the collection, analyses, or interpretation of data; in the writing of the manuscript; or in the decision to publish the results.

References and Notes

1. Muthmann, W.; Kraft, K. Untersuchungen über das Cer und das Lanthan. *Liebigs Ann. Chem.* **1902**, *325*, 261–278. [CrossRef]
2. Porri, L.; Giarrusso, A. Conjugated diene polymerization. In *Comprehensive Polymer Science*; Eastmond, G., Ledwith, A., Russo, S., Sigwalt, P., Eds.; Pergamon: Oxford, UK, 1989; Volume 4, pp. 53–108. ISBN 978-0-08-096701-1.
3. Evans, W.J.; Ulibarri, T.A.; Ziller, J.W. Reactivity of $(C_5Me_5)_2Sm$ and related species with alkenes: Synthesis and structural characterization of a series of organosamarium allyl complexes. *J. Am. Chem. Soc.* **1990**, *112*, 2314–2324. [CrossRef]
4. Taube, R.; Windisch, H.; Maiwald, S. The catalysis of the stereospecific butadiene polymerization by Allyl Nickel and Allyl Lanthanide complexes—A mechanistic comparison. *Macromol. Symp.* **1995**, *89*, 393–409. [CrossRef]
5. Kuran, W. Coordination Polymerisation of conjugated dienes. In *Principles of Coordination Polymerisation*; J. Wiley and Sons: Chichester, UK, 2001; Chapter 5, pp. 275–329.
6. Fischbach, A.; Anwander, R. Rare-Earth Metals and Aluminum Getting Close in Ziegler-Type Organometallics. *Adv. Polym. Sci.* **2006**, *204*, 155–281.
7. Tsutsui, M.; Ely, N. Unusual type of organolanthanide complex containing the allyl moiety. Allylbis(.eta.5-cyclopentadienyl)lanthanide. *J. Am. Chem. Soc.* **1975**, *97*, 3551–3553. [CrossRef]
8. Taube, R.; Sylvester, G. *Applied Homogeneous Catalysis with Organometallic Compounds*; Cornils, B., Herrmann, W.A., Eds.; VCH: Weinheim, Germany, 1996; Chapter 2.3, pp. 280–318.
9. Casely, I.J.; Suh, Y.S.; Ziller, J.W.; Evans, W.J. Formation of a [ONN(allyl)O]−Anion via NO Insertion and Coupling Using Yttrium and Lanthanide Allyl Metallocenes. *Organometallics* **2010**, *29*, 5209–5214. [CrossRef]
10. Fieser, M.E.; Mueller, T.J.; Bates, J.E.; Ziller, J.W.; Furche, F.; Evans, W.J. Differentiating Chemically Similar Lewis Acid Sites in Heterobimetallic Complexes: The Rare-Earth Bridged Hydride $(C_5Me_5)_2Ln(\mu\text{-}H)_2Ln'(C_5Me_5)_2$ and Tuckover Hydride $(C_5Me_5)_2Ln(\mu\text{-}H)(\mu\text{-}\eta^1{:}\eta^5\text{-}CH_2C_5Me_4)Ln'(C_5Me_5)$ Systems. *Organometallics* **2014**, *33*, 3882–3890. [CrossRef]
11. Cui, P.; Spaniol, T.P.; Maron, L.; Okuda, J. Dehydrogenation of Amine-Borane $Me_2NH{\cdot}BH_3$ Catalyzed by a Lanthanum-Hydride Complex. *Chem. Eur. J.* **2013**, *19*, 13437–13444. [CrossRef] [PubMed]
12. Abinet, E.; Spaniol, T.P.; Okuda, J. Olefin Hydrosilylation Catalysts Based on Allyl Bis(phenolato) Complexes of the Early Lanthanides. *Chem. Asian J.* **2011**, *6*, 389–391. [CrossRef] [PubMed]
13. Evans, W.J.; Kozimor, S.A.; Brady, J.C.; Davis, B.L.; Nyce, G.W.; Seibel, C.A.; Ziller, J.W.; Doedens, R.J. Metallocene Allyl Reactivity in the Presence of Alkenes Tethered to Cyclopentadienyl Ligands. *Organometallics* **2005**, *24*, 2269–2278. [CrossRef]
14. Taube, R.; Windisch, H.; Maiwald, S.; Hemling, H.; Schumann, H. XLVIII 1. Synthese und Struktur der ersten neutralen Tris(allyl)lanthanoid-Komplexe $La(\eta^3\text{-}C_3H_5)_3$.1,5 Dioxan und $Nd(\eta^3\text{-}C_3H_5)_3$.Dioxan und ihre Eignung als "single site"-Katalysatoren fur die stereospezifische Butadienpolymerisation. *J. Orgamomet. Chem.* **1996**, *513*, 49–61. [CrossRef]
15. Barbier-Baudry, D.; Bonnet, F.; Dormond, A.; Hafid, A.; Nyassi, A.; Visseaux, M. Organolanthanides, catalysts for specific olefin-diene copolymerization: Access to new materials. *J. Alloys Compd.* **2001**, *323*, 592–596. [CrossRef]
16. Kirillov, E.; Lehmann, C.W.; Razavi, A.; Carpentier, J.-F. Highly Syndiospecific Polymerization of Styrene Catalyzed by Allyl Lanthanide Complexes. *J. Am. Chem. Soc.* **2004**, *126*, 12240–12241. [CrossRef] [PubMed]

17. Kang, X.; Zhou, G.; Wang, X.; Qu, J.; Hou, Z.; Luo, Y. Alkyl Effects on the Chain Initiation Efficiency of Olefin Polymerization by Cationic Half-Sandwich Scandium Catalysts: A DFT Study. *Organometallics* **2016**, *35*, 913–920. [CrossRef]
18. Carpentier, J.-F.; Guillaume, S.; Kirillov, E.; Sarazin, Y. Discrete allyl complexes of group 3 metals and lanthanides. *Comptes Rendus Chim.* **2010**, *13*, 608–625. [CrossRef]
19. Brydson, J.A. Aliphatic Polyolefins other than Polyethylene, and Diene Rubbers. In *Plastics Materials*, 7th ed.; Butterworth-Heinemann: Oxford, UK; Elsevier: Armstrong, The Netherlands, 1999; Chapter 11, pp. 247–310. ISBN 978-0-7506-4132-6.
20. Taube, R.; Windisch, H.; Görlitz, F.H.; Schumann, H. Komplexkatalyse: XL. Darstellung und Kristallstruktur des Tetra(allyl) lanthanat(III)-Komplexes $[Li(\mu\text{-}C_4H_8O_2)_{3/2}][La(\eta^3\text{-}C_3H_5)_4]$, eines Katalysators fur die stereospezifische Butadienpolymerisation. *J. Organomet. Chem.* **1993**, *445*, 85–91. [CrossRef]
21. Robert, D.; Abinet, E.; Spaniol, T.P.; Okuda, J. Cationic Allyl Complexes of the Rare-Earth Metals: Synthesis, Structural Characterization, and 1,3-Butadiene Polymerization Catalysis. *Chem. Eur. J.* **2009**, *15*, 11937–11947. [CrossRef] [PubMed]
22. The lanthanum and the neodymium $Ln(allyl)_2][B(C_6F_5)_4]$ ionic pairs were described and assessed as single-component catalysts for butadiene polymerization but experimental details are lacking, see: Taube, R. Catalytic Reaction Mechanisms and Structure-Reactivity Relationships in the Stereospecific Butadiene Polymerization. In *Metalorganic Catalysts for Synthesis and Polymerization*; Kaminsky, W., Ed.; Springer: Berlin, Germany, 1999; pp. 531–546, ISBN 978-3-642-60178-1.
23. Standfuss, S.; Abinet, E.; Spaniol, T.P.; Okuda, J. Allyl complexes of scandium: Synthesis and structure of neutral, cationic and anionic derivatives. *Chem. Commun.* **2011**, *47*, 11441–11443. [CrossRef] [PubMed]
24. Jian, Z.; Tang, S.; Cui, D. A Lutetium Allyl Complex That Bears a Pyridyl-Functionalized Cyclopentadienyl Ligand: Dual Catalysis on Highly Syndiospecific and *cis*-1,4-Selective (Co)Polymerizations of Styrene and Butadiene. *Chem. Eur. J.* **2010**, *16*, 14007–14015. [CrossRef] [PubMed]
25. Senyek, M.L. Isoprene, polymers. In *Encyclopedia of Polymer Science and Technology*; Wiley: Hoboken, NJ, USA, 2002.
26. Jian, Z.; Cui, D.; Hou, Z.; Li, X. Living catalyzed-chain-growth polymerization and block copolymerization of isoprene by rare-earth metal allyl precursors bearing a constrained-geometry-conformation ligand. *Chem. Commun.* **2010**, *46*, 3022–3024. [CrossRef] [PubMed]
27. Jende, L.N.; Hollfelder, C.O.; Maichle-Mössmer, C.; Anwander, R. Rare-Earth-Metal Allyl Complexes Supported by the [2-(*N,N*-Dimethylamino)ethyl]tetramethylcyclopentadienyl Ligand: Structural Characterization, Reactivity, and Isoprene Polymerization. *Organometallics* **2015**, *34*, 32–41. [CrossRef]
28. Fieser, M.E.; Johnson, C.W.; Bates, J.E.; Ziller, J.W.; Furche, F.; Evans, W.J. Dinitrogen Reduction, Sulfur Reduction, and Isoprene Polymerization via Photochemical Activation of Trivalent Bis(cyclopentadienyl) Rare-Earth-Metal Allyl Complexes. *Organometallics* **2015**, *34*, 4387–4393. [CrossRef]
29. Fadlallah, S.; Terrier, M.; Jones, C.; Roussel, P.; Bonnet, F.; Visseaux, M. Mixed Allyl−Borohydride Lanthanide Complexes: Synthesis of $Ln(BH_4)_2(C_3H_5)(THF)_3$ (Ln = Nd, Sm), Characterization, and Reactivity towards Polymerization. *Organometallics* **2016**, *35*, 456–461. [CrossRef]
30. Fadlallah, S.; Jothieswaran, J.; Capet, F.; Bonnet, F.; Visseaux, M. Mixed Allyl Rare-Earth Borohydride Complexes: Synthesis, Structure, and Application in (Co-)Polymerization Catalysis of Cyclic Esters. *Chem. Eur. J.* **2017**, *23*, 15644–15654. [CrossRef] [PubMed]
31. Rodrigues, A.; Kirillov, E.; Roisnel, T.; Razavi, A.; Vuillemin, B.; Carpentier, J.F. Highly Isospecific Styrene Polymerization Catalyzed by Single-Component Bridged Bis(indenyl) Allyl Yttrium and Neodymium Complexes. *Angew. Chem. Int. Ed.* **2007**, *46*, 7240–7243. [CrossRef] [PubMed]
32. Annunziata, L.; Duc, M.; Carpentier, J.F. Chain Growth Polymerization of Isoprene and Stereoselective IsopreneStyrene Copolymerization Promoted by an ansa-Bis(indenyl)allylYttrium Complex. *Macromolecules* **2011**, *44*, 7158–7166. [CrossRef]
33. Luo, Y.; Chi, S.; Chen, J. Half-sandwich rare-earth-metal derivatives bearing pyrrolidinyl-functionalized cyclopentadienyl ligand: Synthesis, characterization and catalysis in syndiospecific polymerization of styrene. *New J. Chem.* **2013**, *37*, 2675–2682. [CrossRef]
34. Sarazin, Y.; de Frémont, P.; Annunziata, L.; Duc, M.; Carpentier, J.F. Syndio- and Isoselective Coordinative Chain Transfer Polymerization of Styrene Promoted by ansa-Lanthanidocene/ Dialkylmagnesium Systems. *Adv. Synth. Catal.* **2011**, *353*, 1367–1374. [CrossRef]

35. Jian, Z.; Cui, D.; Hou, Z. Rare-Earth-Metal–Hydrocarbyl Complexes Bearing Linked Cyclopentadienyl or Fluorenyl Ligands: Synthesis, Catalyzed Styrene Polymerization, and Structure–Reactivity Relationship. *Chem. Eur. J.* **2012**, *18*, 2674–2684. [CrossRef] [PubMed]
36. Shannon, R.D. Revised Effective Ionic Radii and Systematic Studies of Interatomie Distances in Halides and Chaleogenides. *Acta Cryst.* **1976**, *32*, 751–767. [CrossRef]
37. This represents an activity comparable to the most active titanium and scandium catalysts, see Coates, G.W. Precise Control of Polyolefin Stereochemistry Using Single-Site Metal Catalysts. *Chem. Rev.* **2000**, *100*, 1223–1252.
38. Bonnet, F.; Violante, C.; Roussel, P.; Mortreux, A.; Visseaux, M. Unprecedented dual behaviour of a half-sandwich scandium-based initiator for both highly selective isoprene and styrene polymerisation. *Chem. Commun.* **2009**, *23*, 3380–3382. [CrossRef] [PubMed]
39. Annunziata, L.; Rodrigues, A.S.; Kirillov, E.; Sarazin, Y.; Okuda, J.; Perrin, L.; Maron, L.; Carpentier, J.F. Isoselective Styrene Polymerization Catalyzed by ansa-Bis(indenyl) Allyl Rare Earth Complexes. Stereochemical and Mechanistic Aspects. *Macromolecules* **2011**, *44*, 3312–3322. [CrossRef]
40. Rodrigues, A.-S.; Kirillov, E.; Lehmann, C.W.; Roisnel, T.; Vuillemin, B.; Razavi, A.; Carpentier, J.-F. Allyl *ansa*-Lanthanidocenes: Single-Component, Single-Site Catalysts for Controlled Syndiospecific Styrene and Styrene–Ethylene (Co)Polymerization. *Chem. Eur. J.* **2007**, *13*, 5548–5565. [CrossRef] [PubMed]
41. Perrin, L.; Kirillov, E.; Carpentier, J.F.; Maron, L. DFT Investigation of the Tacticity Control during Styrene Polymerization Catalyzed by Single-Component Allyl *ansa*-Lanthanidocenes $\{(C_5H_4CMe_2(9\text{-}C_{13}H_8)\}Ln(C_3H_5)$. *Macromolecules* **2010**, *43*, 6330–6336. [CrossRef]
42. Kang, X.; Song, Y.; Luo, Y.; Li, G.; Hou, Z.; Qu, J. Computational Studies on Isospecific Polymerization of 1-Hexene Catalyzed by Cationic Rare Earth Metal Alkyl Complex Bearing a C3 iPr-trisox Ligand. *Macromolecules* **2012**, *45*, 640–651. [CrossRef]

MDPI
St. Alban-Anlage 66
4052 Basel
Switzerland
Tel. +41 61 683 77 34
Fax +41 61 302 89 18
www.mdpi.com

Catalysts Editorial Office
E-mail: catalysts@mdpi.com
www.mdpi.com/journal/catalysts

www.ingramcontent.com/pod-product-compliance
Lightning Source LLC
Chambersburg PA
CBHW062106220926
51850CB00010B/2914
* 9 7 8 3 0 3 9 3 6 1 9 0 8 *